BALANCE

BALANCE

*In Search
of the Lost Sense*

Scott McCredie

Little, Brown and Company

New York Boston London

Little, Brown and Company
Hachette Book Group USA
237 Park Avenue, New York, NY 10017
Visit our Web site at www.HachetteBookGroupUSA.com

Library of Congress Cataloging-in-Publication Data

McCredie, Scott.
Balance : in search of the lost sense / Scott McCredie.—1st ed.
   p. cm.
Includes bibliographical references and index.
ISBN 978-0-316-01135-8
1. Equilibrium (Physiology) I. Title.
QP471.M42 2007
612—dc22   2006038089

Q-FF

To my wife, Cher Maillot, who gives my life a better balance, and my parents, Grae and Bill McCredie, who caught me when I fell

Contents

BALANCE

In Search of the Lost Sense

Aristotle was wrong. Often credited as the first to catalog the human body's primary senses, he neglected to mention one of our most important: the sense of balance, that intricate orchestration of nerve impulses that allows us to dance with gravity. His oversight, however, is understandable. While it's easy to detect the role of the eyes, nose, skin, tongue, and ears, the receptors of balance are hidden from view and don't easily give up their secrets. Because the equilibrium sense is mostly autonomic, working below the level of consciousness, like breathing or the beating of a heart, the great philosopher apparently was unable to sense its subtle operation.

Nearly two thousand years later, the American writer Diane Ackerman made a similar error. In her popular book *A Natural History of the Senses,* she glossed over the balance and equilibrium faculties in just a few sentences. Ackerman's short shrift of balance is less defensible than Aristotle's, but it still reflects the common view. For a number of reasons, many people today aren't aware that balance is a legitimate sense. Not only is there disagreement among medical professionals about what constitutes a sense, but balance is a devilishly complicated phenomenon, and to this day it remains half-veiled in mystery and ambiguity.

But among those who have studied the balance sense in depth—specialists in scientific research, physical therapy, and clinical medicine—many would argue that without it we might not have much use for the other senses. Not only is the primary balance organ, called the vestibular apparatus, older than most of the other sense organs on an evolutionary scale—an indication of its importance to the survival of our primordial ancestors—but balance is as vital to our existence today as it ever was. Without balance, many of the things we take for granted would be impossible. We could not stand on two legs, never mind walk or run. We couldn't see images in sharp detail as we move, or navigate without visual landmarks, or perhaps even think clearly.

One reason balance is the "lost" sense is that it was missing in action for so long, hidden in shadows as humans toiled to figure out how their bodies worked. Only in the past 150 years or so have we been able to discover just exactly what it is, how it operates, what can go wrong with it, and how to maintain and improve it. Because it is such a latecomer to the table already occupied by the other senses, balance has had to elbow its way to any sort of recognition. Now it's beginning to get some well-deserved notice. And the timing couldn't be better. Balance is becoming a bigger issue now than at any time in history, mainly because people are "losing" it more frequently than ever before. And when you lose your balance, even for a moment, whether from disease, the effects of aging, or anything else that interrupts your dance with gravity, bad things can happen.

A few years ago my father and I were hiking together on a

popular trail that leads to the top of Mount Si, on the western edge of the Cascade Mountains near Seattle. After stopping at the summit to eat lunch and catch our breath, my dad, who was sixty-seven at the time, grabbed his blue daypack, draped it casually over one shoulder, and began exploring. After a few minutes, he stopped on top of a large rock, and I idly watched him slide the pack strap slowly off his right arm. In what seemed like slow motion, his whole body then proceeded to lean awkwardly toward that side and, to my astonishment, he spun into a headfirst dive, disappearing from sight. Horror-struck, with my heart pounding, I scrambled quickly down to where I imagined him lying, in a crumpled heap, bleeding and badly injured. But by the time I reached him a few seconds later he was already beginning to sit up. Some miracle had caused him to land harmlessly on a patch of ground between boulders. Only his pride had been hurt.

As we descended back down the trail, I began thinking about my father's fall. It was an unsettling demonstration of his growing frailty, and of the dimming of his once dynamic sense of balance. For as long as I had known him he'd been an athletic and graceful man, a renowned fast-pitch softball player, and a professional woodsman who had spent years wading streams and tramping across the slippery backs of fallen logs. I then recalled a friend telling me about a fall his own mother, who was seventy-five, had recently taken. She was a vigorous exerciser who for years had walked several miles at a rapid clip three or four times a week. But on one occasion she suddenly lost her balance and fell forward onto the sidewalk, unable even to get her arms up to protect her face. My friend, though walking at her side, couldn't prevent her fall. The results were gory but not serious.

These two incidents made a deep impression on me and

started me wondering about the nature of balance. What controls it? Is the decline of the human balance system—and falling—an inevitable part of growing old? Why isn't balance as commonly talked about in fitness circles as strength training or aerobics or stretching? I was bemused at first by the lack of information. But the further I looked into the subject, the more intrigued I became, not least because I had to search so far and wide to come up with answers to my questions.

One thing I learned is that balance, like all the other senses, begins to degrade after you reach your thirties—unless the process is forestalled by techniques I'll talk about later. Balance disorders are one of the most common forms of disability in the United States. Although reliable epidemiological data are scarce, a 1989 National Institutes of Health report estimated that some 90 million people over age seventeen "have experienced dizziness or balance problems."[1] This figure includes the broadest range of balance disorders, from serious to mild. For comparison, about 32 million Americans suffer from hearing loss today. In this country, an epidemic of falls is mowing down the elderly like scythes, and soon it will begin thinning the ranks of the used-to-be baby boomers. After the age of sixty-five, one out of three Americans falls each year, adding up to more than 10 million falls annually.[2] Dr. Owen Black, one of the leading clinicians in the field of vestibular disorders, reports that 70 percent of falls are mediated by an impairment of the vestibular system. While most falls don't result in serious injury, they are like a game of Russian roulette. Sooner or later the bullet—in the form of fractures, lacerations, and sometimes death—will sit in the firing chamber. While losing your sense of balance, or any sense, is a predictable part of aging, a number of other diseases and disorders also contribute to the danger, such as ear infections, certain types of brain injuries, diabetes, and the dislodging of tiny stones that reside within the inner ear (yes, every one of us has rocks in our head). Tragically, another source of severe balance loss can afflict otherwise healthy people when they contract a bacterial infection that requires a hospital stay.

Cheryl Schiltz was a vibrant, energetic, thirty-nine-year-old woman living in a small town near Madison, Wisconsin, when she suddenly lost her balance. In 1997, after undergoing routine surgery, she developed a postoperative infection that required the prolonged use of a common antibiotic called gentamicin. Seventeen days later, the drug had succeeded in killing the bacteria. But unbeknownst to her or her doctor, it had also destroyed much of the function of her vestibular system, the pair of tiny bizarre-looking organs nestled within her inner ears. These are the body's dedicated gravity- and motion-sensing organs, similar in function to the gyroscopic guidance systems on a modern airplane. Cheryl's world was soon to be turned—literally—upside down.

Back at home two days after the antibiotic treatment ended, she got up from bed in the morning and, as if struck by a shotgun blast, crumpled to the floor. Despite repeated attempts, she was unable to get to her feet. Her vision was blurry and distorted. After crawling to the bathroom, she managed to slide down the stairs of her two-story home on her rear end. "It was like being incredibly intoxicated," she remembers. "The first thing I thought was: Holy cow, what's going on?"

A short time later, Cheryl was back at her doctor's clinic undergoing tests to determine the cause of her problem. She recalls standing in the corridor—a feat she was able to perform only by bracing herself against a wall—when he delivered the bad news: her balance and vision had been disabled—perhaps permanently—by the effects of the antibiotic. Cheryl was among the small subset of people for whom gentamicin is seriously "ototoxic"—literally ear poison. She was diagnosed with "bilateral vestibular dysfunction," or BVD.

Over the next few years, Cheryl's life proceeded to unravel. She tried going back to work, but her symptoms made that difficult. Driving the hundred-mile daily round-trip commute was nearly impossible due to her blurred vision. It's hard for someone with an intact vestibular system to imagine what it's like to look through the eyes of someone with BVD. "If you were to take a video camera and put it against your chest and just walk around and not pay any attention to what you were doing with it, and then watch the video, that's kind of similar to what it looks like," Cheryl explains. "Things wiggle and bounce around. Your eyes are like they're on springs; they don't want to stay still."

Walking became an act of will. Canes, usually one but sometimes two, were a necessity. As if struggling against the altered gravity of another planet, she had to be constantly vigilant about where she was in space, to *think* about how she was going to get from one point to the next. Assuming what she describes as a "Frankenstein" gait, her body stiff and rigid, her stance wide to compensate for her instability, Cheryl had to keep her gaze focused on the ground in front of her to minimize visual distortion, looking up only to make sure she wasn't going to run into anything. "It's like being in zero gravity," she says. "You just don't have a concept of rightness, or what's up and down. The sense of grounding is gone. It's as if the whole world and everything in it is made out of Jell-O. You don't have anything steady under your feet. And then when you hit that, the Jell-O, everything in the distance starts shaking and that's what you see."

Because she couldn't manage the commute to work, Cheryl was forced to leave the job she loved. She found another closer to home, but her new boss didn't cut her any slack and soon fired her. Although she was a bright, articulate, highly motivated woman by nature, well liked by past employers and colleagues, the BVD not only lowered her self-confidence but diminished many cognitive skills.

"Multitasking was out of the question," she says. "Before the BVD, I'd have two phones going, plus the computer, grabbing this and grabbing that." After the onset of the BVD, "One thing at a time was all I could do. And trying to use a computer, I couldn't take my eyes from the computer to the paper and then back up, it was awful. It would make me feel like I was falling out of my chair. I actually had to get a new chair. I had to ask the company to get me a chair with arms on it, or I would literally fall out of it."

The BVD also affected her short-term memory. She would have trouble remembering information she'd just read. While speaking, words she wanted to say were lost in a kind of brain fog. Simple math became problematic. "I remember making cookies once," she says, "and I got so frustrated, to the point of tears, because I had to go to the other room with the measuring cup and ask my son, 'What is three fourths plus three fourths?' I could not figure it out."

What Cheryl's experience shows us is the utter dependence we humans have on our balance system—not only for maintaining an upright posture, but for vision and even mental acuity. Cheryl, like most of us, never thought about her balance system before it was compromised. Yet we rely on our sense of balance for everything from watching a heron flapping across a marsh to the action (though most people don't think of it as an action) of standing still, which involves constant, nearly imperceptible movements, as well as the integration of many different sensory inputs. Without it we could not perform any of the amazing feats of agility and athletics for which our species is known, from walking a high wire to spinning on ice at eighty revolutions per minute.

Karl Wallenda, the founder of the Flying Wallendas, the famous circus high-wire troupe, possessed arguably the most precisely honed balance of any human in the twentieth century. He came by what most people would consider his almost supernatural balance skills early in life, as many gifted athletes do.[3]

At about the age most kids are learning to ride bicycles, Wallenda was teaching himself to perform handstands in unusual places. It wasn't for sport or his own amusement or because he had nothing better to do. His parents, who had been professional circus performers in Germany, had taught him the rudiments of hand balancing to prepare him for a life in the circus. But when his stepfather was drafted into the German army in World War I, Karl, his mother, and his three siblings were forced to survive on war rations. To supplement the family's skeletal income, the ten-year-old boy, small for his age but agile and strong, would sneak out of the apartment at night, after his mother had gone to bed, and walk alone into town.

Here, at several of the town's restaurants and bars, he would perform a brief but sensational balancing act. First he would do a conventional handstand on the floor, walking around on his hands to arouse the crowd's interest. Then, rising back to a standing position, he would grasp the back of a wooden chair and press up into a handstand. Next he would tilt the chair up so that it was balancing on just its two front legs. The applause would grow a little louder. As the coup de grace, he would then set several chairs on top of one another, balancing all but the bottom chair on two legs. He'd push into a precarious but spectacular handstand on the back of the upper chair, managing to find the column's center of gravity, his feet nearly extending to the ceiling. To magnify the crowd's empathy, he would purposely shift his weight quickly to one side, making it appear that he was about to fall, then recapture his balance. In circus parlance, this bit was known as "selling" the trick to the crowd, making them believe it was extremely difficult. Then he would leap to the ground, landing on his feet, and the coins would fall like sleet into his upturned hat as applause and shouts and whistles engulfed the room.

For Wallenda, whose name would become synonymous with the art of high-wire performing, this entrée was the beginning of a long career spent challenging the limits of human balance. He is an example of how doing so in a methodical way can produce extraordinary results. Even as an old man, he retained supremely good balance, performing regularly through his fifties and sixties, before finally succumbing to gravity at the age of seventy-three. Because of his balance skills, the odds that Wallenda would have fallen the way contemporary older Americans often do were infinitesimally small. Although few health professionals would advocate that their patients learn handstands or wire walking as a way to improve balance, the principles behind Wallenda's approach are the key to unlocking our own mastery of balance. For balance, like aerobic capacity and strength and agility, is highly trainable. And the better your balance, the less likely you are to fall, whether walking down the sidewalk with a cane or slaloming through a mogul field on skis.

This book, then, is an account of my exploration of a subject I've been interested in for years without really knowing it—until the day my father stepped off a boulder and disappeared. In search of the lost sense, I'll follow balance's murky, sometimes quirky trail, unraveling its tangles of complexity, revealing the ways it affects our lives. I'll tell the stories of people like Cheryl who have their balance taken from them, suddenly or gradually, and of those, like circus performers, who have exquisite balance throughout their lives. I'll look at how the balance system can be fooled and profoundly disturbed by forms of movement that it was not designed to register, such as when floating in the zero gravity of outer space, being catapulted off the deck of an aircraft carrier, or even riding in the backseat of a car. I'll show how our balance system contributed to the survival of the human species, from increasing our hunting agility to helping us keep from getting lost. I'll examine the question of why it took so long for us to fathom balance as a sense, and how bizarre spinning machines for quelling unruly lunatics were the predecessors of tools used by a diverse and distinguished group of nineteenth-century scientists to unlock the secrets of balance. I'll pursue the somewhat mysterious connection between balance and cognition, analyzing how some people appear able to sharpen their mental skills by honing their balance system. Along the way I'll discuss techniques, both simple and sophisticated, for training, maintaining, and enhancing balance, to help people keep dancing right on through to the last song. By the final chapter I hope you'll understand why I believe that balance—this marvelous, almost unbelievably complicated thing our bodies do—is so important to our well-being that it demands to be elevated to its rightful position among the other five senses. For in the end, balance may prove to be the most primary—as in primordial, life-sustaining, essential—of all the senses.

Chapter One

Sickness from Motion

Ho Goddess Nausea!...
'Neath whose ruthless power
Valour and virtue, strength and genius cower.
    — FROM "ODE TO SEA SICKNESS," 1884,
    BY WILLIAM MUIR

Though the boat ride from hell happened decades ago, the memory lingers like a vile odor. When I was sixteen a friend invited me to accompany him and his family on a salmon-fishing trip along the British Columbia coast. All twelve of us were crammed into a twenty-foot runabout, its vinyl cover stretched over the normally open cockpit, sheltering us from the summer drizzle. As we reached an open channel, the boat met a big swell from the southeast, locking itself into a relentless, rhythmic motion. Sitting upon the inboard motor housing at the stern of the craft, facing forward, I soon began to feel sweaty and lethargic. The subtle queasiness in my stomach gradually turned into full-blown nausea. I remember staring straight ahead, glum and miserable, unable even to look out through the scratched, nearly opaque plastic windows to the gray monotony beyond. The smell of exhaust fumes permeated every breath. I was thinking about how and where I was going to heave my lunch without attracting too much attention when I heard a voice. "You'll never guess what color your face is," chirped John, grinning like a ten-year-old tying tin cans to a cat's tail. "Oooh, it's sort of green!" someone else remarked.

I had succumbed, not for the first time and not for the last, to what is perhaps the most common form of balance disorder: motion sickness. Though it strikes about a third of people who travel by air, sea, or land, motion sickness can be induced in almost anyone with an intact vestibular system, depending on the type and duration of the stimulus. On a scale of severity, it ranks at the bottom of the dozen or so known balance-related disorders. For most people, the symptoms go away soon after the stimulus ends, and it's usually not life-threatening, though many victims, I can attest, feel as if they would rather die. It's a common adage among experienced sailors that when you're sick from the sea, first you're afraid you're going to die, and later you're afraid you won't.

After that first memorable bout, I became vaguely aware, as most people are, that motion sickness was somehow related to an inner-ear problem, but I had no clue how the whole thing worked. It turns out that until the latter part of the nineteenth century, nobody else did either. Even today, though millions of dollars have been spent researching it, controversy still exists about the fundamental nature of motion sickness, how best to treat it, and how it might have served our species, in an evolutionary sense. "It seems extraordinary that...the fundamental nature of the disorder is still not understood," wrote researchers in a study published in the *Annals of the New York Academy of Sciences* in 1992. "In particular," the authors noted, "it is not clear why it exists at all."

Motion sickness is a curious phenomenon. For one thing, it's really not a "sickness" in the strict sense of the word. There are no bacterial or viral invaders, no misfunctioning organs. It's a common and normal response of the body to certain well-defined movements, but it can also occur, as we shall later see, when the body is completely still. Age and sex seem to play a role in susceptibility. Infants under two years old rarely experience it, while children between two and twelve are especially prone.[1] A 1955 study of five thousand passengers of transatlantic troopships showed that 31 percent of seventeen- to nineteen-year-olds reported seasickness, while just 13 percent of thirty- to thirty-nine-year-olds did.[2] After age fifty, motion sickness is rare,[3] unless, for instance, you happen to find yourself on a life raft in rough seas, where virtually 100 percent of people become seasick.[4] Apparently due to female-specific hormones, women are "more susceptible than men to all forms of motion sickness."[5]

It would be natural to assume that motion sickness is a disorder of the stomach, because that's where the nausea seems to originate. But it's actually a sensory problem. The sense involved is not one of the five we learn about as children, but that elusive sixth: our sense of balance.

Unlike the other major senses, our sense of balance doesn't make use of just one kind of sensory organ. You use your eyes for sight, your nose for smelling, and so forth, but balance employs a varying ratio of three different sensory inputs: vision, proprioception (cells in muscles and joints that indicate their position in space), and the vestibular system. Motion sickness involves two of the three sensory components of balance, vision and the vestibular system.

While most people have some familiarity with how the eyes function, they can't say the same of the vestibular system, also known as the vestibular apparatus, labyrinth, or simply the inner ear. Yet it is one of the most wondrous organs in the body, one that connects us across eons to some of the oldest and most "primitive" animals on earth, and which has played a crucial role in our development as a species. There are two of these labyrinths, located within the inner ear, each about the size of a marble. They look like a science fiction writer's concept of creatures from another galaxy who have taken up residence inside our skulls. If they *were* aliens, you would guess that they were quite intelligent by the location in which they had chosen to live. *Vestibule* is a Latin term that means

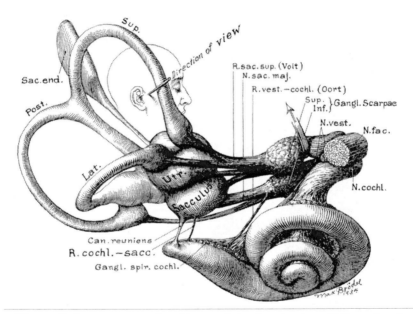

The vestibular apparatus and cochlea are shown here in a beautifully rendered drawing made in 1934 by Max Brödel. On the left side are the three "rings" of the semicircular canals, labeled "Post." for posterior, "Lat." for lateral, and "Sup." for superior. The "Utr." and "Sacculus" represent the utricle and saccule. The snail-like appendage is the cochlea. (Original art in the Max Brödel Archives, Department of Art as Applied to Medicine, The Johns Hopkins University School of Medicine. http://www.hopkinsmedicine.org/medart/history/ Archives.html)

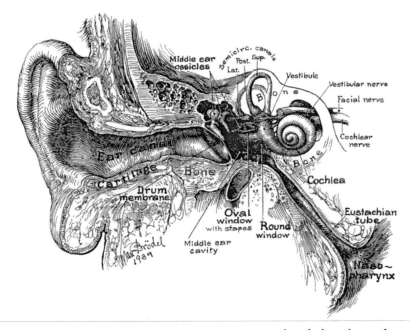

This drawing, done by Max Brödel in 1939, gives an idea of where the vestibular apparatus lies within the inner ear. You can see the semicircular canals, the vestibule, and the cochlea in the upper right, protected by a bulge of very hard bone. (Original art in the Max Brödel Archives, Department of Art as Applied to Medicine, The Johns Hopkins University School of Medicine. http://www .hopkinsmedicine.org/medart/history/Archives.html)

"entryway" or "antechamber," originally used in architecture to describe an area between the door and the interior parts of a house. In human anatomy, it refers to the entryway into the skull, which here is fortified like a castle, surrounded by some of the hardest bone in the body, the "petrous," or stony, section of the temporal bone.

The "head" of the vestibular creature—three heads, actually—is formed by the semicircular canals, which look like three tiny hollow rings linked together. Their mazelike structure accounts for the name *labyrinth*. Attached is the cochlea, a major component of the hearing system, which curls upon itself like the grasping tail of a sea horse. Because we live in a three-dimensional world, the three semicircular canals are oriented approximately along the three dimensions of space, which, as most of us learned in high school math, correspond to the x, y, and z axes of a graph. When a person's head rotates, fluid moves in the canals, exciting minute hair cells. Those cells send signals to the brain that tell it in what direction, how fast, and how far the head is moving.

The other extraordinary parts of the labyrinth are called the utricle and the saccule. Nestled directly beneath the semicircular canals, these two saclike structures house hundreds of tiny particles of calcium carbonate, a chemical compound that is the primary component of limestone, marble, snail shells, blackboard chalk, and antacids such as Tums. These are called *otoliths*, a Greek word that literally means "ear rocks." In a scanning electron micrograph, otoliths appear as sharply faceted crystals jumbled together like boulders at the bottom of a landslide. The weight of this "heavy" mass of otoliths makes the utricle and saccule sensitive to linear (straight-line) accelerations of the head and to the pull of gravity. These organs are activated when, for instance, you roar down a runway in an airplane or while beginning an ascent inside an elevator (both linear movements), or when you tilt your head forward (gravitational force). The otoliths lend mass to the gelatinous material they're embedded in. So when the head moves in a straight line or tilts in any direction, the crystals cause underlying hair cells to bend, in turn sending signals to the brain telling it which direction the head is moving in, or, if the head is tilted, what position it's in relative to gravity.

The vestibular apparatus plays one of the most important roles in the balance system. It constantly measures our body's position in space, relative to gravity, and is the only organ in the body dedicated exclusively to balance functions. But by what means does this amazing organ, under certain conditions, reduce us to pathetic, wimpering souls?

Shortly before Aristotle trumpeted the five human senses, another Greek, Hippocrates, took notice of the nauseating effects of travel by ship. In his treatise "The Nature of Man," he wrote that "sailing on the sea shows that motion disorders the body." In fact, the word *nausea* is derived from the Greek word for ship: *naus*. As citizens of a powerful seafaring nation, the ancient Greeks knew a thing or two about seasickness.

I was in good company when it came to historic figures known to have suffered from mal de mer, as the French call it. Julius Caesar, Charles Darwin, and, most surprisingly, Lord Nelson, the famous British admiral and revered war hero of the late eighteenth century, all suffered badly and often from seasickness.[6] Nelson, according to historical accounts, was seasick practically every time he set sail, especially during the early days of his career, but also during the last voyages of his life. Another Brit, Lawrence of Arabia, rode camels the way Nelson sailed ships of the line and was often sickened by their lumpy, rolling, undulating gait. When Napoleon suggested that a dromedary corps be formed in Egypt, to serve as scouts for his troops, the idea was quickly squelched when it was learned that riding camels caused motion sickness. A research study in 1970 found that camel riders frequently complained of the disorder, as did those who traveled by elephant, but equestrians were never bothered by a horse's motion.

Humans aren't the only vertebrates who can become ill from motion. Two British scientists reported in 1975 that "dogs are about as susceptible as man; other less susceptible animals are horses, cows, monkeys, chimpanzees, seals, some birds—and even some species of fish get seasick, when they are put in a tank on a boat and transported over the rough surface of the water."[7]

For more than two thousand years after Hippocrates identified seasickness, the origin of the disease was unknown, though dozens of hypotheses attempted to explain it. The variety of alleged causes demonstrates just how all at sea these learned men were in trying to fathom nausea caused by motion. In the nineteenth century, one argument pinned the blame on parts of the body that moved independently and were thus susceptible to passive movements. Shaking of the liver was one example, in which it was thought that bile discharged into the small bowel caused nausea. Likewise, some doctors thought that when the contents of the stomach were jostled by passive motion, of a ship for instance, a reflexive action was initiated that led to vomiting. Others believed that either too much or too little blood in the brain, stomach, or chest was responsible for motion sickness.

Nausea provoked by motion, far from being simply a condition to treat, was itself considered a useful treatment for some patients. In the early nineteenth century, psychiatrists dealt with certain mental conditions by placing patients in rapidly rotating chairs, which induced vomiting and supposedly increased beneficial blood flow to the brain. With this same goal in mind, physicians later in the century prescribed sea voyages for some patients, not for fresh air or a change of scenery, but for the express purpose of making them seasick. Perhaps these doctors were simply following the ancient guidance of several Greek philosophers, who believed that motion sickness was at least a partial remedy for "consumption, insanity, dropsies, tumours, apoplexy, elephantiasis, and 'many diseases of the head, breast, and eyes.'"[8]

Another alleged cause of motion sickness, amazingly, was sound. One Boston doctor advised using wax to plug the ears, citing as evidence the experience of Ulysses and the Sirens. Other experts believed disturbed vision was to blame, or salt water finding its way into the lungs, or some kind of shock to the nervous system. So many educated guesses, over so many centuries, failed to find the source of this common malady.

This abundance of possible causes of the disorder created a blank canvas for "cures." There were nearly as many remedies as ocean swells, from the mundane to the absurd, from benign to lethal. Some of the more quaint remedies called for herbs in various combinations, such as this concoction from a treatise published in 1607 by an anonymous English doctor:

If in your drinke you mingle Rew with Sage
All poyson is expel'd by power of those,
Who would not be sea-sick when seas do rage
Sage-water drink with wine before he goes.

Sir Francis Bacon, in the seventeenth century, tells the story of a "certain English-Man, who, when he went to sea, carried a Bagge of Saffron next to his Stomach, that he might conceale it, and so escape custome; And whereas he was wont to be always exceeding sea-sick; At that time he continued very well, and felt no provocation to vomit."

In the nineteenth century, according to J. T. Reason and J. J. Brand's 1975 classic, *Motion Sickness,* physicians dispensed cures based on their pet notions of the disorder's origin. In the 1870s, doctors frequently prescribed amyl nitrate, a potent chemical compound that dilates blood vessels and increases heart rate, to combat what was thought to be "blood congestion" in the brain and spinal cord. On the other hand, those doctors who believed in the "gut shift" theory often recommended some sort of belt or girdle to be worn around the waist, to slow down the movement of stomach contents. Similarly, there was little agreement on the proper diet for someone who might encounter rough seas. Some advised dry toast, while others thought more exotic foods were preferable, such as "soup made of horse radish and rice, seasoned with red herrings and sardines." Reason and Brand were particularly taken by the dietary prescription of one nineteenth-century "expert" who said passengers "should consume pickled onions prior to embarkation so that the resulting gas might distend the stomach and thus maintain a steady pressure within the abdominal cavity." When the ship got to rolling wildly, passengers were supposed to hold their breath and contract their stomach muscles. "After a pre-embarkation diet of pickled onions," Reason and Brand wrote, "it is possible that this course of action might have led to some social embarrassment!"

Ingenious mechanical remedies were attempted, including variations on the hammock. Since Columbus, British and French sailors had adopted the use of hammock beds from Central and South American natives. Hammocks not only saved space in cramped quarters but, presumably, could counteract the rocking and rolling movements of a ship. In fact, Charles Darwin, while sailing aboard the *Beagle,* often sought refuge in them when he was struck by mal de mer. "The misery I endured from sea sickness," he wrote, "is far beyond what I ever guessed at. The real misery only begins when you are so exhausted that a little exertion makes a feeling of faintness come on — I found that nothing but lying in my hammock did any good." Specially built hammocks were sometimes set up for passengers in the midships area (the part of the vessel that experienced the least amount of rolling), as well as "articulated cabins or couches" that probably used a kind of gimbal arrangement to maintain an upright position.[9]

These inventions were the forerunners, in a sense, of modern guidance mechanisms for airplanes and rockets that allow pilots to know their orientation in space—upside down, right side up—even when they can't see.

Headway toward understanding the true nature of seasickness finally came about in the late nineteenth century. In 1881, two scientists, Irwin and de Champeaux, published separate papers that pointed out striking similarities between the symptoms of seasickness and those of a peculiar disease described twenty years earlier by a French doctor, Prosper Ménière.[10] Ménière had shown that a disturbance of the vestibular system was the source of an ailment later named after him: Ménière's disease. Because the symptoms of seasickness and Ménière's disease—pallor, dizziness, sweating, and vomiting—were nearly identical, the root cause was probably the same, Irwin and de Champeaux hypothesized. Further evidence came a year later, according to a 1916 article in the *New York Medical Journal,*[11] when a man referred to only as "W. James" made an astute observation while sailing aboard a vessel rocked by heavy weather. On board was a group of fifteen deaf mutes. While many people aboard the ship fell ill, James observed that none of the deaf mutes did, which struck him as odd. Perhaps the deaf mutes' damaged inner ears had something to do with their immunity to seasickness, he reasoned. Ménière and others had recently discovered that the inner ear was not only involved in hearing, but, via the vestibular apparatus, had a major role in controlling equilibrium. Was it possible that a normally functioning vestibular system was required to become seasick?

The investigator's full name, it turned out, was William James, father of American psychology, philosopher, and brother of the novelist Henry James. In 1874, William James, trained as a physician and physiologist, began teaching psychology at Harvard University and founded the nation's first psychology laboratory, though he had never taken any courses in the subject himself. James's influence on the profession was profound; soon after he began teaching at Harvard, dozens of other universities started their own programs, a professional society was formed, and several psychology journals were published. Paradoxically, though he was a great advocate of experimental psychology, James himself was loath to spend time in the lab. Even so, he felt compelled to organize an ambitious experiment to test his hypothesis about deaf mutes and seasickness. The preliminary report was published in the *Harvard University Bulletin* in 1882, and a century later it was detailed in his posthumous book *Essays in Psychology.*

After making inquiries, James discovered that no one before him had ever noticed that deaf mutes were often immune to vertigo and dizziness, "another illustration," he wrote, "of how few facts 'experience' will discover unless some prior interest, born of theory, is already awakened in the mind." He arranged with several deaf-mute asylums to examine some of their inmates. With help from his brother Henry, James and his colleagues tested more than 500 deaf-mute children and adults. The subjects were rotated either while standing, eyes closed, or while seated on a square wooden plank, suspended by ropes. The results "beautifully" confirmed his hypothesis: "A very large number of the deaf-mutes examined are either wholly incapable of being made dizzy by the most violent rotations, or experience but a slight and transient giddiness." Of the 519 subjects, 36 percent were not dizzy at all, while 26 percent experienced slight dizziness and 38 percent exhibited normal sensitivity to the rotations. In the general population, most people, usually over 90 percent, would have become disturbingly dizzy under these conditions. (For all its simplicity, this experiment produced reliable results; in 2004, Swedish scientists used state-of-the-art technology to test the inner-ear function of a group of deaf mutes and came up with findings nearly identical to those of James.)

James didn't stop with the spinning tests. He also wondered how deaf mutes would respond in an environment where gravity was ambiguous. Would an impaired inner ear allow them to determine their spatial orientation in such conditions? To find out, he conceived of an experiment perhaps most notable for its cruelty. In a large pool of water, he dunked the deaf mutes one at a time beneath the surface with their eyes closed. What James witnessed made a profound impression on him:

> Every one who has lost himself in the woods, or wakened in the darkness of the night to find the relation of his bed's position relative to the doors and windows of his room forgotten, knows the altogether peculiar discomfort and anxiety of such "disorientation" in the horizontal plane...Imagine a person without even the sense of gravity to guide him, and the "disorientation" ought to be complete—a sort of bewilderment concerning his relations to his environment in all three dimensions will ensue, to which ordinary life offers absolutely no parallel. Now this case seems realized when a non-dizzy deaf-mute dives under water with his eyes closed. He hears nothing (except perhaps subjective roaring); sees nothing; his semicircular canal sense tells him nothing of motion up or down, right or left, or round about; the water presses on his skin equally in each direction; he is literally cut off from all knowledge of their relations to outer space, and ought to suffer the maximum possible degree of bewilderment to which in his mundane life a creature can attain.

Why would this experience be so terrifying? If a person with a normally functioning vestibular system were dunked, he would at least have some idea of how his body was positioned and oriented in the water. Even though he couldn't see, or sense his position via proprioception (because gravitational cues are absent underwater), he could still sense movements of his head and body through the vestibular system and thus have at least a rough approximation of his position. But take away the vestibular system, and there would be complete and utter disorientation. He would literally not know up from down, left from right, and he would feel like an astronaut untethered from his craft during a space walk, tumbling uncontrollably across the void. We can only imagine the howls of terror or anger some of the deaf mutes, emerging from the pool, must have unleashed on James and his assistants.

Thus seasickness was the nausea and vertigo caused by "over-excitement" of healthy, fully functioning semicircular canals, according to James. His conclusions about the role of the vestibular system on spatial orientation presaged ideas that would develop a century later about the possible role of the inner ear in dead reckoning, a navigational skill some animals, including humans, use to find their way in the world. His experiments proved prescient in another way: a century later, one of the caveats always given to people with malfunctioning vestibular systems is to be extremely cautious when swimming or diving because of the danger of underwater disorientation.

Inspired by James's work with deaf mutes, another researcher, a German named A. Kreidl, devised an experiment to confirm the American's hypothesis.[12] He built a platform that could be manipulated to re-create the effects of a rolling sea. Various animals were placed on the platform until they became seasick. Kreidl then surgically severed the animals' eighth cranial nerve, the one that carries signals from the vestibular system to the brain. Again, he set the animals on the platform and tried to make them sick—to no avail.

So by the start of the twentieth century, it was fairly well established that an agitated, healthy vestibular system was the primary component of seasickness. A 1916 *New York Times* article on the subject carried the headline: "Cause of Seasickness Discovered at Last?"[13] It went on to outline the research carried out by scientists in the previous fifty years that led up to the "absolute knowledge" of the true cause of seasickness and vertigo. There were no "epoch-making" discoveries responsible for the knowledge, the author stated, but a series of small contributions. He described the question of seasickness as "a bewildering and complicated matter, and the solution of these mysteries a notable achievement in scientific research." But the "solution" wasn't as pat as the writer implied; it would be another twenty or thirty years before all the pieces of the puzzle would fit together because, unknown to scientists at that time, several pieces were missing from the box.

Although most scientists agreed on the origins of the disorder by the time the *Times* article was published, many of the old explanations persisted until World War II. That's when a wave of new research was conducted by governments concerned with protecting their armed forces from the debilitating, and thus potentially lethal, effects of seasickness. Soldiers reeling from nausea after a few hours asea in a landing craft would not, when they hit the beach, be in top fighting form, nor would bomber crews or men transported in troop ships across long stretches of ocean.

Before the war, no reliable relief had yet been found to comfort those afflicted by motion sickness, though many remedies had been suggested. They ranged from amyl nitrite, potassium bromide, and creosote, to cayenne pepper, opium, cocaine, and "Valerianic ether of menthol on a sugar lump."[14] One scientist noted that claims for most of the various medicines were unconvincing, as none had been tested in a rigorous way. Another researcher, after looking at all the unsubstantiated claims for motion sickness "remedies," drolly stated that he was "convinced that the only real and complete cure is to sit under a big tree."[15]

So the race was on, in Britain, Australia, Canada, and the United States, to find treatments more practical than seeking refuge in a forest. The British and Canadians embarked on a program that focused on swinging test subjects in a special chair, like the ones James used to analyze deaf mutes, to determine how individuals reacted to "nauseogenic" motion. For those who were only moderately susceptible to motion sickness, there was some evidence that they could become used to it, like a sailor who gets his sea legs a week or two into the voyage. The idea was to eliminate people who were most susceptible and to see if the rest, through repeated exposure, could adapt. But both notions proved impracticable; the first because it would screen out too many men at a time when the demand for soldiers was high, and the second because it took too long to train a man to withstand motion sickness. The American research program also looked at what kinds of forces provoked motion sickness most. Instead of using a swing, scientists placed subjects inside a cab that moved up and down in a specially built eighteen-foot-long elevator shaft, in which the variables of acceleration, velocity, and amplitude could be carefully controlled. The most "nauseogenic" or sickness-inducing type of motion, it was discovered, occurred at a frequency range of about one cycle every three to four seconds. (You can form a mental picture of this by imagining yourself in a boat with a wave passing beneath you every three to four seconds.)

Despite the best attempts of several governments, no explanation was found for differences in susceptibility to motion sickness. Reason and Brand, the authors of *Motion Sickness,* cited several studies that looked at the percentage of people who became nauseous after certain types of stimulus, but concluded that "we must content ourselves with the broad statement that *all* individuals possessing an intact vestibular apparatus *can* be made motion sick given the right quality and quantity of provocative stimulation, although there are wide and consistent individual differences in the degree of susceptibility." What caused these individual differences was mysterious in the 1940s, and remains so today.

British and Canadian scientists, abandoning their screening efforts, conducted extensive trials to test various drug treatments to combat motion sickness. The clear winners were those whose primary component was an extract of the belladonna plant, also known as nightshade, deadly nightshade, and devil's cherries. The name *belladonna* is Italian and means "fair lady," a reference to women who once used the plant to enhance their beauty. One of belladonna's well-known properties, first mentioned by the Greek physician Galen in the second century, is that it dilates the pupils, making the eyes appear dark and, presumably, seductively large. (It is still used today by ophthalmologists for dilation.) Belladonna also gives the skin an attractive pale cast, like today's makeup foundation. But if a woman didn't want to wind up in the morgue, the dose needed to be carefully measured, for belladonna is extraordinarily poisonous to humans. It's in the same family as tomatoes, peppers, and potatoes, but it is said that people have been poisoned simply by eating the flesh of game that have fed on its berries, which grow as large as cherries. Symptoms of poisoning include not only dilation of the pupils, but also flushed skin—the opposite of its desired effect among Italian beauties—and dryness of the mouth and throat, sometimes to the point of vocal impairment. In high doses, belladonna can cause hallucinations so vivid that "the very fabric of reality will break down," one experienced user wrote. "You can be sitting down watching TV one moment and the next see your dead grandmother next to you on the sofa asking for more tea."[16] The user's concept of space and movement may also be dramatically altered. People have reported episodes of flying or falling through the air, completely free of gravity. Perhaps most puzzling is that belladonna poisoning also causes vomiting and nausea, the last things one might expect from a remedy for seasickness.

But three of belladonna's chemical components explain why the plant has been prescribed extensively, in medicine as well as in "magical" concoctions, for centuries: the alkaloids atropine, scopolamine, and hyoscyamine. Atropine, now synthetically produced, is a staple of every hospital intensive care unit in the United States. It's a first-line drug used when a patient's heart begins to beat dangerously slowly. It quickly increases the heart's contraction rate and force, and shrinks the diameter of blood vessels (increasing blood pressure). Scopolamine, and to a lesser extent hyoscyamine, decrease acid secretions in the stomach and reduce nausea. Scopolamine is the active ingredient today of Transderm Scop, the little round, flesh-colored patches people put behind their ears before a cruise or fishing trip. Scopolamine acts to block nerve signals from the vestibular system to the brain. Even sixty-five years after it was first introduced to alleviate seasickness, the chemical has few peers: the company that manufactures it claims that over a million prescriptions for Transderm Scop have been written since it came on the market.

In 1944, the Allies issued seasickness capsules containing belladonna-derived alkaloids to troops marshaling for the Normandy invasion. Seas in the English Channel just before the attack were extremely rough, and some men wound up taking more than the prescribed dose of anti-seasickness pills. The soldiers who accompanied Ernie Pyle, the revered American journalist, were among them. Pyle later wrote that "by noon all the Army personnel aboard were in a drugged stupor. The capsules not only put us to sleep, but they constricted our throats, made our mouths bone dry, and dilated our pupils until we could hardly see. When we recovered from this insidious jag, along toward evening we threw all our seasickness remedy away, and after that we felt fine."[17] Pyle reported, however, that the pills did succeed in preventing seasickness in himself and all the men.

In the postwar years, motion sickness research waned. There was still no unified theory about precisely how or why it occurred, despite all the investigations into exactly what motions caused it and the best ways to prevent it. One major advancement during this period was an accidental discovery, in 1947, that led to the introduction of a new class of drugs for reducing symptoms. It happened this way: The manufacturer of a new antihistamine gave the drug to the Allergy Clinic at Johns Hopkins University to study its effects on hay fever and hives. Doctors gave it to a pregnant woman who suffered from hives. But when she started taking the drug, she noticed an odd side effect: she no longer felt sick when riding in an automobile or streetcar, a problem that had plagued her all her life. After she reported the observation to her doctor, the drug was tested in clinical trials to see if other people had the same reaction. A troop transport was enlisted during its ten-day crossing of the Atlantic from New York to Bremerhaven, Germany. Seasickness struck men given the antihistamine far less frequently than those who received placebos. The air force then conducted its own tests, using planes whose pilots simulated turbulent air with nauseating combinations of rolls, yaws, and pitches. After twelve flights, it was determined that

28.7 percent of the men who had taken the drug became ill, in contrast to 55.6 percent of the placebo group. Similar results were found in a test of car sickness. The drug was Dramamine, now a well-known anti–motion sickness remedy.[18] Somehow it altered the central nervous system in a way that prevented motion-sickness nausea. But again, scientists weren't sure exactly how it worked. Even today the mystery persists.

When the space race between the two cold war superpowers began in the early 1960s, suddenly there was a renewed urgency to investigate motion sickness. It became apparent early on that space flight caused the same sort of motion sickness that existed on Earth. But the consequences were far more dangerous. Focused, unfettered minds were vital to making sound judgments at a moment's notice. And barf in one's space suit, especially during a space walk, not only was messy and difficult to clean up, but could wreak havoc with life-support systems or cause death through aspiration.

The Russians were plagued with space sickness in their first forays into orbit, while the Americans avoided it through most of the Mercury and Gemini flights. But that changed during the Apollo program. By 1971, it was reported that ten of the twenty-one Apollo astronauts had experienced space sickness of varying intensities, from "mild sensations of tumbling" to full-on nausea and vomiting. The general public wasn't made aware of this problem, owing perhaps to the macho image of the larger-than-life astronauts. In fact, some of the astronauts even attempted to hide their conditions from colleagues on the ground. Don Parker, a wiry, white-haired vestibular researcher who worked for NASA during the early 1970s, recalled an incident aboard the second Skylab mission that let the cat out of the bag. "On all the Skylab flights," Parker said, the hint of a smile playing on his face, "there was an open mike, so anything that was said was recorded and eventually transcribed. My memory of it is one of the astronauts said, 'Throw it out the chute and we won't tell anybody.'" A couple of days later, somebody on the ground came across the transcript and called up to Skylab to ask what was in the ejected container. "It was the first of many barf bags," Parker continued. "He had been really, really sick. It turned out that there were a whole bunch of people who were having considerable problems with space motion sickness. A friend of mine who was an astronaut gathered information regarding space adaptation syndrome [noting the frequency of space sickness]. And the head of the astronaut office talked about him as a traitor."

Another incident of motion sickness that created a great deal of anxiety at Mission Control occurred on the *Apollo 9* moon mission. Two of the three astronauts reported symptoms of space sickness, and Russell Schweickart was particularly hard-hit. The mild nausea he experienced at the start of the flight only got worse as the moon loomed larger. When it came time for him to slide into his space suit and pilot the Lunar Module, he vomited. NASA had no choice but to delay the mission. Hours ticked by as Schweickart took it easy, keeping his movements to a minimum, allowing the antinausea drugs to kick in. After finally recovering, he performed his duties flawlessly.

Researchers later speculated about why the Russians and Americans had such different experiences with space sickness during the early flights. They found that if an astronaut was free to move his head around inside the spacecraft, he was more likely to be sick. The first two Russian craft were larger, allowing more room to move inside the cabin, than either the *Mercury* or *Gemini* capsules. The *Apollo* vehicle, however, was about the same size as the early Russian craft, and its crew reported a similar rate of space sickness. Reducing head movement appeared to diminish motion sickness, in a space capsule or, it turns out, in a pitching boat or gyrating airplane.

Even with the widespread use of highly effective anti-motion sickness treatments—both belladonna derivatives and antihistamines—there were still occasions when all such measures proved useless, when the brain could be overwhelmed by nauseogenic stimuli. Space flight was one example. But even hardened navy crews aboard ships at sea were susceptible under extreme conditions. Bill Brinkman, a U.S. Navy submarine electrician's mate in the early 1960s, described a voyage that made my salmon-fishing trip sound salubrious by comparison.[19] His vessel, the USS *Sea Owl*, had been participating in NATO exercises in the Caribbean off Bermuda when Hurricane Gracie reared up and canceled the operation. While attempting to outrun the storm, the sub became caught in its jaws. You'd think a sub could simply submerge to get out of the weather. But in a storm like that, which might last longer than the diesel-powered sub could remain underwater, it was considered safer to stay on the surface. If, while below, the sub had to get to the surface in an emergency, say because of a fire, there was a possibility that it would emerge broadside to the waves and capsize. So the *Sea Owl* ran smack into the fury of hurricane-tossed waves up to fifty feet high. Brinkman reported that of the hundred crewmen, all but ten became seasick. He must have been one of the few who remained healthy, for other crew members offered him $100 to stand their four-hour watches so they could remain in their bunks. On the fifth day, the *Sea Owl* entered the eye of the hurricane, an oasis of calm seas and clear blue skies. "It was wonderful," Brinkman remembered. "It was as though we were in a huge soup bowl 20 miles in diameter. We were in...extremely smooth water, as smooth as a still lake, and all around...was a rim of water 50 feet high [and a] swirling black cloud which you could not see through." While cruising through the eye, many of the ninety sick sailors regained their health and were even able to eat. But it was a cruel trick. As soon as they passed through the eye and back into the fifty-foot seas, the men got even sicker than they had been before.

Submarines, and in fact any vessel where the passengers or crew cannot see outside—my friend's runabout with the enclosed canopy is a good example—are probably the worst sorts of craft to be on when the seas are nauseogenic. Why? To answer this question, the theory of what causes motion sickness had to evolve beyond 1940s beliefs.

During World War II, an Australian researcher named McIntyre made what proved to be a prescient observation. He took William James's idea that motion sickness has a vestibular origin and went several steps further. McIntyre noticed that people could be made motion sick simply through "unusual visual stimuli alone."[20] Such things as wearing glasses with the wrong prescription or watching flickering or rapidly moving objects could make someone sick even when there was no unusual motion to stimulate the vestibular system. What was going on here? How could there be sickness without movement? Moreover, he found that flight crews got sicker if they could not see the horizon, say when flying in clouds or sitting in areas of the plane with no windows. McIntyre was one of the first to intuit that when a conflict arose between what the eyes saw and what the vestibular system sensed, motion sickness could be provoked even without motion.

With this observation, he threw his weight behind the so-called sensory conflict theory. Others had talked about this idea before, even as far back as 1881. But only in the thirty years after the war would it become firmly established in the scientific community. To understand how it works, let's examine the case of bomber flight crews with limited vision. The crew's vestibular systems, sensing the jostling and motion of the plane, told them they were moving, while their eyes, seeing only the stationary interior of the fuselage, said they were still. Pilots were much less likely than navigators to get motion sick because they were able to look outside and see that they were moving. In addition, pilots were in control of the craft and thus could anticipate movements of the plane. Meanwhile, navigators watching their instruments or poring over maps experienced great conflict between signals from their eyes and inner ears, a perfect recipe for motion sickness.

This same situation occurs on many family car vacations—the driver rarely becomes motion sick, while kids reading in the backseat often do. McIntyre was perhaps the first to notice the simplest way to alleviate motion sickness in planes, cars, or boats. "Visual impressions during manoeuvres," he wrote, "are the best means of re-orientation and suppression of the unusual vestibular sensations." In other words, stop reading that book and look out the window! Or, in my case on the boat, if it hadn't been raining that day and we'd cruised without the vinyl cover fastened over the cockpit, the sensory conflict might have been reduced enough to spare me the dreaded mal.

Later, the theory of sensory conflict was expanded to include not only conflict between senses, but conflict within a single sense—between what the brain views as "normal" and "abnormal" based on past experiences. This phenomenon was termed "exposure history." The concept was first investigated by a scientist named Stratton in 1897, who "wore an optical system that both inverted and reversed the retinal image while leaving the inputs to the vestibular and non-vestibular proprioceptors unchanged."[21] Stratton's normal vision was rearranged by the experimental lens, and the ensuing nausea was caused by the mismatch between what the brain expected to see and what it actually saw. Exposure history may also explain why babies seem to be immune to motion sickness. In the first several months of their lives nearly all movement they experience is passive. Consequently, their vestibular systems wouldn't perceive a mismatch when subjected to the sort of passive motion that an older child or adult would find nauseogenic.

Until the 1960s, despite McIntyre's research and the corroboration of other scientists, the prevailing idea among motion sickness experts was still that the disorder was caused by overstimulation of the vestibular system. Some scientists tried to explain away nausea induced by visual stimuli alone as a separate disorder. But the symptoms were too similar to dismiss.

Another argument in favor of the sensory conflict model was the mysterious malady called *mal de débarquement*. In French, this term means literally "sickness from getting off a boat." Given enough time, most people eventually adapt to nauseogenic environments. The classic example is sailors in the eighteenth or nineteenth century who spent months or years at a time on oceangoing vessels. They usually got their sea legs after a few weeks (except, of course, for the unfortunate Lord Nelson). For most people, the brain adapts at all levels—visually, proprioceptively, and within the vestibular system—so that what used to be considered unusual stimuli eventually become normal. But some people, mostly women in their forties and fifties who have returned from at least a weeklong cruise on a ship, feel a distinct rocking sensation after they step off the boat.[22] While many weekend sailors often experience something like this when they return to shore and, walking along the dock, still feel the movement of the boat as they regain their land legs, this feeling usually goes away after an hour or two. In people with *mal de débarquement,* however, the brain doesn't readapt to land for days or, in rare instances, months and even years. None of the sensory systems is being overstimulated, yet the unsteady feeling persists, though almost never with nausea or other symptoms of motion sickness. The only way to explain this difficulty, scientists say, is that a conflict has arisen between the new condition—stable ground—and the unstable shipboard environment to which the brain has adapted. But the exact cause of the disorder remains unclear.

People suffering from *mal de débarquement* are often frustrated when they try to seek medical help. Most family practice doctors aren't familiar with the disorder and often dismiss the symptoms as having a purely psychological origin. One woman reported that she was sent to a mental ward, where a psychiatrist, believing she was depressed, ordered her to undergo shock treatment.[23] Somewhat surprisingly, the therapy succeeded in eliminating her symptoms for eleven days afterward. But it hasn't become part of standard treatment protocol.

In fact, motion sickness itself has often been viewed as at least partly a psychological disorder, and there is some evidence for this, both anecdotal and clinical.

In one study in 1964, carried out for NASA, two groups of ten people were rotated in a device designed to make them motion sick. In one group, the subjects were asked to make mental calculations, solving mechanical and spatial problems.

The other group was told just to close their eyes and relax. Twice as many people became nauseated in the eyes-closed group (six) as in the brain-exercising group (three).[24]

Something of a Pavlovian response is evident in the story told by one navy submariner. He had a tendency to get seasick when the sub cruised on the surface but of course was fine underwater, where the seas were always smooth. It got to the point where every time the command was given to surface, his stomach would start to feel queasy. Motion sickness researchers report the same dynamic in some of their test subjects: greater sickness occurs in those who anticipate becoming sick.

The *Journal of Applied Psychology* published an Israeli study in 1995 that looked at seasickness and the power of self-fulfilling prophecy, also referred to as a "verbal placebo." Two groups of naval cadets were about to embark on a five-day training cruise. One group was told the conditions would not be likely to cause seasickness and that, even if they did, the cadets' performance would not suffer. The control group was told nothing. At the end of the cruise, those in the first group reported less seasickness than the control group, and their performances were rated higher. So theoretically, based on this study, if someone had told me a few hours or days before the start of my fateful runabout voyage that I wouldn't get sick, I might have been okay — or at least less green.

The placebo effect may also be at work in one of the more popular remedies for seasickness, wristbands that apply subtle pressure or electrostimulation to an acupuncture point on the underside of the wrist called P6. Sold under the brand names Acuband and ReliefBand, they have been the subject of several controlled trials for the relief of motion sickness, but the results have not been consistent. Part of the attraction of alternative remedies such as acupressure wristbands, as well as gingerroot and slow, rhythmic breathing, is that (1) they're not drugs, for those who don't like to take them, and (2) they have none of the side effects that plague most of the effective drug treatments.

Robert Stern is professor emeritus of psychology at Pennsylvania State University and a NASA researcher on motion sickness. With a high forehead accentuated by a crown of baldness, he has dark, arched eyebrows and a winning smile. In discussing the alkaloid scopolamine, he says, "We've shown in my lab that small doses increase normal gastric activity, and this is probably one of the factors that promotes its anti–motion sickness quality. It also depresses central nervous system activity, tranquilizing some people and, perhaps, making them less aware of minor symptoms of motion sickness. On the downside, it causes dry mouth, drowsiness, and sometimes blurred vision." Scopolamine, as mentioned before, dilates the pupils, which is probably what causes vision to blur. Many of the antihistamine drugs, of which Dramamine was the first, can cause dizziness, drowsiness, and respiratory problems. "Dramamine," Stern continues, "relieves some of the symptoms of motion sickness. Some people get relief from the dizziness, others get relief from the nausea. We really don't know the mechanism of action. The most serious side effect is sedation. That's why antihistamines are included in some over-the-counter sleeping medications. When astronauts are given an antihistamine to prevent space motion sickness, they are usually given a stimulant along with it to prevent drowsiness."

The basic forms of these first-line antimotion drugs were discovered fifty years ago, with only modest refinements since then—a startling fact given how much pharmaceuticals, at least in some arenas, have advanced since then. Asked about the slow pace, Stern replies, "Why aren't there better drugs?

Probably because the greater the relief from symptoms of motion sickness, the greater the negative side effects. And no one ever died from motion sickness."

So far, the only viable, proven, drug-free method of alleviating the symptoms of motion sickness is to allow the brain enough time to adapt. Most pilots, sailors, and astronauts, after enduring a few uncomfortable hours or days, will eventually stop getting air, sea, or space sick. It's as though the brain builds a stockpile of motion memories that eventually becomes large enough to accommodate any newly encountered movements. As experience banks fill with different sorts of movement, there's less of a chance for conflict to occur between how the brain registers past and present motion, which may explain why the incidence of motion sickness begins to decline rapidly after puberty and usually disappears beyond age fifty.

One study that beautifully demonstrates the concept of adaptation was published in *Physical Therapy* in 1999. It focused on a thirty-four-year-old marine biologist who had complained of extreme sensitivity to motion sickness, triggered when driving a car, when riding in an elevator, and, worst of all because it affected her job performance, while standing on a floating dock or scuba diving. The researchers had her perform a series of "visual-vestibular" and balance-training exercises at home, progressing in stages over a period of ten weeks. One exercise was to hold an index card, printed with half-inch-high letters, at arm's length, moving it back and forth and up and down while focusing on the letters. At the beginning of the therapy, this movement would incite nausea after only thirty seconds. After ten weeks, she could tolerate it without difficulty. Another exercise focused on standing balance. Before the therapy, she had been unable to walk on a foam surface (or a floating dock) without having to sidestep to keep from falling. The therapy started with marching in place with her eyes closed, then progressed to walking on foam cushions, and finally to doing everyday household activities while wearing special foam boots. At the end of the study, she was not only able to navigate the foam cushions with complete control, but also could comfortably walk on floating docks. She reported no dizziness when driving and only mild motion sickness after scuba diving for three hours. "Though she was not completely free of symptoms in the most provocative conditions," the researchers wrote, "her ability to function in these situations was no longer limited and symptoms were mild." The habituation appeared to last long after the exercises stopped. Ten months later, the marine biologist said she was still able to function well at work and at home.

Virtual-reality environments are the latest frontier in experimenting with adaptation to motion sickness. A significant percentage of people become nauseous after playing virtual-reality games and viewing simulations. This phenomenon is closely related to the sickness-evoking effects of flight simulators and Cinerama movies. A mismatch occurs between what the eyes see (motion) and what the vestibular system senses (no motion). Scientists in the Visual Ergonomics Research Group at Loughborough University, in Leicestershire, England, created an experiment to look at habituation to a virtual environment. They picked nineteen subjects to view a popular video game called Wipeout, wherein futuristic hovercraft glide around a track at high speed. The game, chosen for its "particularly powerful nauseogenic stimulus," was viewed on "head-mounted displays," also known as personal display systems. Miniature screens sit in front of the eyes like a pair of oversized glasses. After twenty-minute viewing sessions over five consecutive days, the participants reported less nausea on day five than they had on day one, and the researchers concluded that most of the change was due to habituation.

But adaptation, of course, doesn't do much good for people who want to prevent motion sickness from striking in the first place—landlubbers about to embark on a weeklong sailing voyage, for instance, or astronauts preparing for a shuttle flight. With about 50 percent of Space Shuttle passengers falling prey to motion sickness (and 75 percent on their first flight), NASA is highly aware of the problem and continually searching for ways to soften the blow. In the absence of gravity, the inner ears' otolith organs don't function properly because there's nothing to tell them where "down" is. Most shuttle astronauts say they feel "upside-down" when they're in space because of this ambiguity. The signals the otoliths send to the brain are in conflict with input from the semicircular canals, and motion sickness often results.

So how do astronauts, who have access to state-of-the-art technology, deal with space sickness today? Dr. Stern says that they receive more or less the same kinds of drugs that you or I would take before a rough boat ride. The only difference would be that when an astronaut takes an antihistamine to lessen the nausea, he'll also down a stimulant to prevent drowsiness. There are, as of yet, no magic bullets for motion sickness.

As the human species lopes into the future, experts say that we will inevitably be exposed to more and more experiences that will cause motion sickness. In Europe, trains that "tilt" in corners, like a banked roadway, to achieve higher speeds with greater passenger comfort, are becoming more common. Yet up to a third of the passengers report motion sickness, especially if they can see outside. It's the opposite of reading-in-the-backseat syndrome, but still adheres to the same principle:

a conflict between vision ("yikes, we're tilting!") and the vestibular and proprioceptive senses ("everything feels normal"). In the future, we'll probably tap into more virtual-reality environments, to play games, learn how to fly or drive, watch movies, or perhaps even reduce our fear of spiders. (One application now being studied has subjects view virtual-reality simulations of big hairy spiders crawling up their arms, which eventually reduces fear through habituation.) And though it's probably still a long way off, perhaps our everyday transportation will one day resemble the hovercraft in Wipeout. Then we may all be wiped out by bouts of motion sickness.

Science *is* making headway toward alleviating motion sickness in the virtual world. One of the leaders in this field is the University of Washington's Human Interface Technology Laboratory in Seattle. That's where Don Parker, the former NASA scientist and vestibular system expert, now does his research. When he was working for NASA, Parker developed a trainer that allowed astronauts to preadapt to weightlessness and space sickness. His NASA grant permitted him to go anywhere in the world, and he chose the University of Washington. But after NASA decided that medication was more feasible than adaptation training, Parker's grant was canceled. So he got involved with friends who worked in the UW's HIT Lab, where he began work on anti–motion sickness techniques for the virtual environment. He and his colleagues came up with four different methods, one of which was patented. Independent visual background, as they called it, consisted of placing in the background an image (clouds, for example) that doesn't move as the rest of the scene does, but rather matches the orientation and motion detected by the viewer's vestibular system. So instead of having a conflict between vision and vestibular sensory inputs, "what we do is give you a component of the visual scene that does match your inner ear," Parker explained. For instance, in a fixed-base simulator, like a driving simulator, where the virtual world moves and the subject is stationary, the clouds would remain motionless in the background. This simple adjustment eliminates much of the discomfort caused by motion sickness, according to Parker.

Yet the question remains: Just why do humans, and a few select species, react so revulsively to certain types of movement or perceived movement? What quirk of evolution allowed this sensory hypersensitivity to persist? Scientists have pointed out that in certain circumstances motion sickness may decrease one's chances for survival, as in the case of pilots and sailors who wind up in life rafts on a stormy sea. It's been reported that almost all of them succumb to motion sickness, which probably hastens their demise because vomiting rapidly increases dehydration. There is some speculation that motion sickness is an "evolutionary anomaly," whose preliminary symptoms, such as blurry vision and dizziness, just happen to mimic those of stomach poisoning. When the brain perceives those signals, it thinks, I'm being poisoned, it's time to purge. That sensory confusion wouldn't have been a problem for the thousands of years humans moved about on foot. But our bodies, specifically our vestibular systems, were simply not designed to endure the passive motion imposed on us by boats, cars, trains, planes—or camels. They, along with computer monitors, IMAX screens, and virtual-reality displays, seem to disturb our brain's view of the way the world ought to be.

Chapter Two

Van Gogh's Ear

As the eyes are windows to the soul, so disease is a portal through which science can fathom mysteries of the human body. When a system or organ goes awry, its secrets are often revealed. And people who experience the malady may become conscious, perhaps for the first time, not only of an organ's existence but of its role in their ability to live a normal life. That is certainly the case with the human vestibular system.

In the autumn of 2004, Irene Langston didn't know she had a balance disorder when she walked into the MultiCare Vestibular and Balance Clinic, in Tacoma, Washington. She knew only that her balance was bad, and that her family practice doctor, unable to find the cause, had sent her to the clinic for testing and evaluation. A sparkly ninety-two-year-old with a quick smile, Irene clung to a walker and shuffled along with unsteady steps as she made her way to a chair. She had been ordered by her doctor to use the walker since having back surgery two years before. In the meantime, she'd also had a hip replaced and undergone quadruple bypass heart surgery. Her balance problem, though, had a longer history.

One night in 1987, she had fallen after getting out of bed to go to the bathroom, and had to call 911 for help. Later her doctor told her the fall had been caused by an inner-ear infection known as labyrinthitis, a condition usually caused by a virus. (The infection can generate inflammation of the vestibular system, which may falsely signal the brain that motion is occurring, inducing dizziness.) But although the ear infection disappeared, Irene's equilibrium was never quite the same. She'd be okay for two or three months, then experience a dizzy spell, usually when she was lying down in bed at night. Lately, the spells had gotten worse, and may have contributed to the fall that broke her hip.

Now she was sitting across from a physical therapist, Karen Perz, and me. I had come to the clinic to learn about some of the ways the balance system can go awry, and how a physical therapist treats these problems. Although there are a slew of medical specialties that deal with diseases of the vestibular system, physical therapists who specialize in balance deal most directly with the ways people manage to live with, and overcome, their maladies. When I had spoken with Karen on the phone, she suggested I sit in with her one day and observe. Arriving at the clinic, I was struck by its *Romper Room* qualities: one large area held a treadmill, stationary bike, balance beam, and exercise balls, and in a back room stood a diagnostic machine that resembled a telephone booth. Karen was in her midforties, dark-haired, with a no-nonsense demeanor and playful wit.

Like a highway patrolman giving a sobriety test, Karen first checked Irene's balance by having her stand up and close her eyes, which she did without trouble, and then walk a straight line heel to toe. Next Karen asked Irene to stand on one leg, which she did for about a nanosecond. Karen then had Irene lie on a padded platform that resembled a double bed.

"What I want to do," Karen said, "is see where the dizziness is coming from. I'm going to have you lie down and see what your eyes do." Karen then proceeded to use a special test developed in the early 1950s by a famous British otologist, Charles Skinner Hallpike, and his assistant, Margaret Dix: the Hallpike-Dix maneuver. After Irene removed her glasses, Karen helped her lie down on her left side and cradled Irene's head in her lap. Irene looked up and Karen peered into her eyes. "That's interesting," Karen said, "they're dancing and dancing."

I wasn't sure what she meant until she had Irene switch sides and invited me over to look. What I saw was amazing. Irene's eyes were glassy and blue and, just as Karen had described, dancing in their sockets in rapid, circular, jerky motions. It was as though she were possessed by some alien being, like Linda Blair in *The Exorcist*.

What Karen had provoked was a phenomenon called nystagmus. The eyes jerk involuntarily back and forth in their sockets, which is similar to what happens when you're riding a train and looking at the view out a side window. As things catch your attention, a bull in a field for instance, your eyes focus on it for a few moments, following the image as the train moves along. When the bull leaves your field of vision, your eyes quickly jerk forward to fix on another object, a colt perhaps, and follow it for a few milliseconds, and so on. In fact, there is a name for just this sort of eye movement: railroad nystagmus. It isn't truly nystagmus, however, because it's voluntary. There are about fifty different names that describe variations of nystagmus, which is derived from the Greek word *nustagmos*, which means "drowsiness." The association between sleep and nystagmus may have to do with the rapid eye movements that can sometimes be observed in sleeping people.

The particular kind of nystagmus that Irene experienced is called vestibular nystagmus. It happens because there is an intimate connection between the vestibular system and the eyes called the vestibular ocular reflex, or VOR. This reflex allows the eyes to stay focused on an object when the head moves sideways. When our heads turn, say, to the left, a signal travels from the vestibular system to the eyes to tell them to move slightly to the right. This tiny movement of the eyes is enough to compensate for the head movement and keep the object focused on the retina; otherwise, it would appear blurry, like the view looking through a movie camera panning quickly from side to side. You can demonstrate to yourself just how useful this mechanism is by holding your palm about a foot in front of your eyes. Now shake your head back and forth at the rate of about three times a second and focus on your palm. You'll probably be able to see even fine details; everything remains in focus. That's because your vestibular system is unconsciously moving your eyes at the precise speed necessary to counteract every head movement. Now try keeping your head stationary and moving your hand back and forth in front of your eyes, again at the rate of about three times a second. No matter how hard you try to focus on your palm, the details blur out.

Just as the balance system itself is "invisible" to us, so too the VOR works without our awareness. But the VOR can be tricked into operating when there is no head movement. That occurs, for instance, if while standing you pirouette rapidly several times and then suddenly stop. The world seems to spin for several seconds afterward, and you may feel dizzy. If a friend were to look into your eyes at that moment, he would see them dancing in their sockets, though *you* would have no idea they were moving. The receptors of your inner ear are still telling your eyes that you're rotating, and your eyes try to compensate for the movement with VOR. What happened with Irene is that as Karen laid her on her side and moved her head in a certain way, her inner ear believed, falsely, that she was rotating. It then gave off the signal to her eyes to go into nystagmus and caused her to feel dizzy.

Karen knew at once that the particular pattern of nystagmus in Irene's case was associated with one of the most common balance disorders among people fifty and over, responsible for at least half the cases of dizziness in this age group.[1] The long name for it is benign paroxysmal positional vertigo, but it's often referred to simply as positional vertigo. It is considered "benign" because it isn't life-threatening—unless it strikes when the sufferer is standing atop a ladder, for example, or scaling a mountain. "Paroxysmal" means it happens occasionally, at odd intervals. And "positional" indicates that it is triggered by certain, usually predictable, positions of the head. That's why it is sometimes called "top shelf" vertigo; when a person tilts the head back to look up, the symptoms may begin. For Irene, and many others who have positional vertigo, the episodes are triggered while in a horizontal position, either when getting out of bed or when rolling over.

What's going on within her vestibular system that sparks the vertigo? While the condition has been recognized for eighty-five years, it was only in about 1960 that a probable mechanism for it was found. The most widely accepted theory is that minute particles of calcium carbonate, each about three microns wide, which reside within the inner ear, become dislodged. They then migrate into another section of the inner ear, inadvertently stimulating nerve receptor cells.[2] Those cells erroneously tell the brain that the head is being rotated, which then causes dizziness, vertigo, and nausea.

How the calcium particles become dislocated is something of a mystery still. Although a head injury or viral infection can loosen them, in about half the cases there is no known cause. For some people, it appears simply to be part of the aging process.

Karen wants to see if she can position Irene's head in such a way as to coax the errant calcium particles away from the portion of her inner ear where they're causing problems. She's going to perform a movement similar to the Hallpike-Dix maneuver. This one is officially called the canalith respositioning procedure, though it's commonly referred to as the Epley maneuver, a nod to the Portland, Oregon, otologist, John Epley, who invented it in 1979. When Karen Perz, who's been a physical therapist since 1983, switched to specializing in vestibular and balance disorders in 1991, she had never heard of the Epley maneuver. Perhaps that was because the technique was still considered only a hypothesis then. Evidence for its efficacy was circumstantial. The difficulty of observing the workings of the vestibular system prevented scientists from actually seeing the particles as they were being repositioned. In a study published in 1992, Epley wrote: "Although it cannot be directly shown that [the maneuver] acts through actually mobilizing [calcium particles] out of the semicircular canal as envisioned, the fact that carrying out a procedure designed with that object has proved very effective in control of [positional vertigo] lends support to this premise." Karen says that when Epley first went public with it, after years of using the technique on his own patients, it wasn't well received until several years later, after research had been done that backed his claims of success. Now it is standard clinical practice around the world.

So once again Karen asks Irene to lie down on her side and cradles her head. "When we do this," Karen says to her, "it will make you dizzy. I'm telling you this so that you're not caught off-guard. If you begin to feel like you might throw up, tell me." Irene assures her that she hardly ever gets nauseous.

Seconds after her head is gently guided to the right, however, Irene says in a soft voice, "I'm feeling dizzy." A moment later, with more urgency, she says she's getting sick. I hand over a wastebasket and she vomits into it.

Then Karen helps her into a sitting position and puts her arm around her. "You're probably one of five or six people who have [vomited] in the fifteen years I've been doing this," Karen tells her. "It probably doesn't feel very good to be in this distinct category, does it?"

Irene laughs meekly.

"The reason people get sick," Karen explains, "is because it's like being seasick."

"I used to go out fishing," Irene replies, "and I never, ever, ever got sick."

Because nausea often occurs when the vestibular system is disordered, disturbed, or stimulated in a certain way, whether from motion sickness or a variety of other conditions, Karen had managed to ferret out a major clue toward identifying Irene's problem by provoking symptoms in Irene that resembled seasickness. Though she probably didn't appreciate being made dizzy and sick, Irene would soon enjoy the fruits of Karen's detective work. Before Irene is out the building, Karen is on the phone with her family practice doctor. She explains what she's discovered and asks the doctor to write a prescription for an anti–motion sickness drug that Irene will take before coming in for her next visit. Free of motion sickness, Irene has a very good chance of being cured by the Epley maneuver after two or three visits. After enduring spells of dizziness and vertigo off and on for sixteen years, Irene had at last found somebody who could help her.

Many family practice doctors are unfamiliar with balance disorders, so patients who suffer from them are often forced to live with their dizziness and vertigo. Although dizziness is one of the most common complaints of people who go to the doctor, it's an enormously complicated problem to figure out, with dozens of possible causes. Some are related to neurological disorders, some are caused by blood pressure problems, and some are due to inner-ear disturbances. In the case of positional vertigo, Karen tells me, people diagnosed before about 1990 were told that nothing could be done—which was true back then. Most of them simply stopped complaining about it. One patient of Karen's lived with positional vertigo for twenty-five years, unable to find anyone who knew enough about the problem either to help her or to refer her to someone who could. And she was not a typical patient. A well-educated woman who worked as a middle manager in a hospital, she was someone who might be expected to know about the latest developments in balance disorders. In the future, the number of people with undiagnosed positional vertigo should decrease, as information about the disorder spreads. And with the growth of balance-oriented physical therapy like Karen performs, more and more people like Irene will regain the sense the rest of us take for granted.

Nausea wasn't the first symptom Robin Grindstaff, another of Karen's patients, felt after an errant mortar shell exploded fifty feet away from her during a military training exercise. That would come a few years later. The accident, in Germany in 1981, severely injured both Robin's ears, destroying 30 percent of her hearing. Army surgeons rebuilt the damaged sections in a series of surgeries that allowed her to hear well enough to avoid hearing aids. But a few years later, the intricate repair failed, her hearing deteriorated, and she began having attacks of vertigo. She clearly remembers the first one.

"I was standing in a parade formation," Robin says, "and all of a sudden it hit, and I fell. They pulled me back behind a tree and said, 'Maybe you're dehydrated.' Two weeks later I had another [attack]. And then I thought, Okay, something's not right. They were coming pretty regular. Last anywhere from five minutes to a couple of hours. One or two a week. And then when they diagnosed it they said, 'Hmmm, there's not a lot we can do for those.'"

For someone who's never experienced one before, it may be hard to imagine how debilitating a severe vertigo attack can be, though the National Institutes of Health report that 40 percent of the population will, at one time or another, experience vertigo disturbing enough to seek a doctor's help. Terence Cawthorne, a British otologist (a physician who specializes in the vestibular system), wrote in 1945 that the experience is "often so terrifying in [its] intensity that observers unused to the ways of the labyrinth may find it difficult to believe that such a profound disturbance can be caused by injury to such a modest organ. The overwhelming vertigo, the awful sickness and the turbulent eye movements—all enhanced by the slightest movement of the head—combine to form a picture of helpless misery that has few parallels in the whole field of injury and disease."

Most of us have at least some experience of vertigo from our youth, a memory of staying too long on a playground merry-go-round or enduring one of those carnival "fun" rides. The world seems to spin crazily around you, or you feel as if you're wildly gyrating in space. Walking, let alone standing, becomes almost impossible. If you're not hanging on to something, you'll probably fall. In most people, this feeling goes away after a few minutes, but for people with vestibular disorders, like Irene and Robin, these attacks can last for several hours and strike without warning.

Robin, who was stationed at Fort Lewis, in Washington State, when vertigo first struck, was sent to an ear, nose, and throat specialist, an otolaryngologist, to find out what was causing the problem. The doctor performed a series of tests that determined she had something called Ménière's disease, a condition characterized by an excess of fluid in the inner ear. When the pressure builds to a certain level, the fluid stimulates receptors that send false signals to the brain that the head is rotating. In a sense, the extra fluid acts like the dislodged calcium particles that cause positional vertigo. Classic signs of the ailment are vertigo, ringing in the ears (called tinnitus), a feeling of fullness or pressure in the ear, and fluctuating hearing loss.

Though most cases of Ménière's don't have identifiable triggers, it was obvious that Robin's was probably caused by the shell blast, as head injury is one of the known sources of the condition. Other suspected causes of Ménière's are autoimmune disorders, allergies, and viruses, and there's probably a genetic component as well. It is an exceedingly rare disease, striking one in five hundred, most predominantly in the forty-to sixty-year-old age group.[3]

Vincent van Gogh was probably in his early thirties when he allegedly contracted Ménière's disease. The Dutch painter, renowned for "making pictures," as he described it, as well as for his bizarre and turbulent life, has been the subject of endless analyses by modern-day medical sleuths. All have tried to discover what the strange affliction was that drove him to self-mutilation and suicide. In 1986, four American doctors presented a paper at the Second International Symposium and Workshops on Surgery of the Inner Ear entitled "Vincent's Violent Vertigo." Later published in the *Journal of the American Medical Association,* the paper, by I. K. Arenberg and others, argued that van Gogh suffered not from epilepsy, the diagnosis given by his doctor and the one most frequently cited by contemporary art historians, but from Ménière's. The authors, three of whom worked for the International Ménière's Disease Research Institute in Englewood, Colorado, were perhaps a little biased in their approach, and the very nature of their research—sifting through 796 letters that van Gogh wrote to relatives and friends to hunt for the artist's descriptions of his symptoms—seems imprecise at best as a way to pin down an accurate diagnosis. Still, their conclusions, though controversial and conjectural, shed light on how Ménière's can transfigure a life.

Van Gogh, who lived from 1853 to 1890, noted his first attacks of dizziness and vertigo when he was living in Paris in the 1880s. "In Paris...I was always feeling dizzy...and at that time it was recurring to me rather regularly," he wrote in a letter to his sister. He describes "attacks" of debilitating "vertige" (which Arenberg and his coauthors translate as vertigo) that make him "feel a coward before the pain and suffering." The attacks lasted longer than epileptic seizures usually do, and were followed by months-long periods of calm, as Ménière's attacks often are. The authors also contend that "nowhere in his 796 extant letters does he describe a characteristic convulsive seizure of the grand mal type...no bitten tongue or post-seizure oral bleeding, or muscle fatigue; no involuntary self-inflicted injuries from muscle flailing." Further, during the year he was a patient at the St. Rémy Lunatic Asylum—he voluntarily committed himself, after eighty citizens of Arles, his hometown, signed a petition requesting that he be sent there—he drew no comparisons between his symptoms and those of the epileptics around him.

Epilepsy, a disease of the nervous system usually characterized by "fits" or "seizures," was poorly understood in van Gogh's time. It was then considered a form of insanity and was often, according to an early-twentieth-century physician the authors cite, a "wastebasket term for a vast array of central (and peripheral) nervous system disorders." A Frenchman who lived and worked in Paris at the same time van Gogh did, Prosper Ménière was, in 1861, the first to describe the malady later named for him. But his discovery wasn't accepted by the medical community for many years, and in fact few physicians outside of Paris knew of his work. Arenberg and his coauthors argue that van Gogh's doctors, like most medical men of his era, were not familiar with Ménière's disease and so made the misdiagnosis of epilepsy, some of whose symptoms are similar to Ménière's.

To support their hypothesis, the researchers point to a 1979 paper by a Japanese otologist named Yasuda, who also believed that van Gogh suffered from Ménière's disease. Yasuda, apparently, made even greater leaps of imagination. He asserted that the very style of van Gogh's painting was influenced by the nystagmus experienced during his attacks, quoting the artist's description of "vertical tremors" to support his claim. Yasuda even discerned a visible change in van Gogh's style that, he argued, was dependent on whether he was in the grips of an attack (which could last for months). As to the famous ear-severing incident, could van Gogh have felt so much pressure in his ear, and perhaps so much characteristic tinnitus, that he was driven to cut it off? For some reason the *JAMA* authors shy away from making that argument directly, though they include Yasuda's claim that he did and reinforce the position by stating: "We have encountered patients who have said that tinnitus and/or aural pressure were so severe that they would 'cut off their ears' or 'poke a hole (into their ear) with an ice pick' if this would free them of their symptoms."

Although Ménière's usually affects just one inner ear, between 25 and 50 percent of patients, within five years, suffer impairment in both ears. In Robin's case, both of her inner ears are affected (called bilateral Ménière's). Karen Perz tells me she rarely sees such a severe form of the disease.

Robin was able to stay in the army through most of the 1990s despite her illness, but her doctors didn't really have any solutions for stopping her sudden vertigo attacks, other than prescribing diuretics to prevent buildups of inner-ear fluid. This protocol was the standard of the day, as there were few medical answers for anyone suffering from Ménière's at the time. Robin had a noncombat job working in Fort Lewis's personnel center, keeping track of soldiers as they arrived and departed, handling promotions, inspecting facilities, and instructing. In 1993, when the United States sent troops to fight in the war in Somalia, Robin's outfit was summoned. She was recovering from one of her many ear surgeries when the call came, but her commander didn't believe her condition should keep her home.

Soon after arriving in Africa, Robin spent an eight-hour day working outside in searing heat. Despite liberal applications of sunscreen, she got severely sunburned. Her skin blistered so badly that some came off as she removed her shirt. Medics told her to stay out of the sun for at least the next week, but her company commander once again decided she shouldn't get special treatment. He sent her outside again the next day. After becoming extremely dehydrated, she went into kidney failure, a life-threatening condition. "When you're on diuretics," Robin explained, "your potassium levels have to be monitored really close because you can die if they go too high and go into kidney failure if they go too low. My doctors had always told me to drink orange juice and eat bananas to put potassium back in your system, but when you're in the desert you don't get those things."

She had to be medevaced back to Fort Lewis, after doctors stabilized her at a hospital in Germany. When she was back in the States, her Ménière's symptoms kicked in more forcefully than ever. The vertigo attacks increased and her hearing worsened. Doctors decided to try an experimental surgery, inserting a shunt into her inner ear that drained excess fluid down to her spinal column. It didn't work.

By 1997, her balance had become so bad that she was ordered not to practice on the firing range, for fear she might fall with a loaded weapon in her hands. In 1999 she retired from the military and, after a stint in her home state of Michigan, moved with her husband and three children back to Tacoma, near Fort Lewis.

When I met Robin in Karen's office in 2004, I was impressed by her determination and lightheartedness, though later I would realize that her humor masked the tremendous sadness she feels about the way the disease has shaped her life. A short woman, a little overweight, with grayish blond hair, she occasionally uses a cane decorated with small American flags. Although she appears to be able to hear normally, she actually has very little hearing ability, even with her two hearing aids. In fact her hearing has been bad for so long that she has developed an almost uncanny ability to read lips. Karen always waits for Robin to look at her before she speaks, so that Robin can watch her lips move. On the day of our first meeting, Robin wore blue jeans and a gray sweatshirt emblazoned with the words PIERCE COLLEGE, the community college she attends. As Karen directed her through a series of exercises designed to retrain her sense of balance, she seemed a little distracted and nervous. Karen later told me that in a few hours Robin was going to undergo yet another surgical procedure that was supposed to diminish the symptoms of her Ménière's disease. In the process, however, doctors were going to take away the last vestige of her hearing. She knew that by the weekend she would be completely deaf.

The type of Ménière's that Robin has is the most severe and difficult to treat. It's not just that it affects both of her inner ears, but that each is affected differently. "One side was saying 'Stand up!' and the other side was saying 'Fall down!'" Robin says, describing the mismatch. After a great deal of testing, doctors determined that the disease was worse in her left ear. So a month before I met her, they decided to deactivate her left inner ear, removing one source of the faulty balance signals. The brain, which has a marvelous way of adapting to new sensory input (but only if it is stable) would then learn to interpret the faulty signals from the right inner ear. Even though erratic, the information would be more constant.

That was the hope, anyway. A lot of things can go wrong in this kind of surgery, which involves cutting the vestibular portion (as opposed to the hearing portion) of the eighth cranial nerve, the conduit of all sensory information to the brain from the ears. The trouble is that both portions are intertwined and enmeshed. Not wanting to entirely destroy her hearing, the surgeon didn't slice through the whole nerve but dissected it as carefully as he could. Apparently, he didn't sever enough of the vestibular portion. Though a little hearing still remained, so did some of the vestibular function. Robin's vertigo, after disappearing for a week after the surgery, came back stronger than ever, along with severe head pain.

To destroy the last remaining nerve cells of the vestibular system, doctors decided the best approach was to inject the antibiotic gentamicin—the same antibiotic that accidentally caused Cheryl Schiltz's balance disorder—into Robin's inner ear. The chemical is commonly used in this procedure because of its known destructive qualities to the inner ear. It is thought that most people, given the gentamicin regimen that Cheryl underwent after her surgery, don't have their inner ears destroyed. Cheryl was unlucky, one of a very small percentage of people who have this reaction, perhaps for genetic reasons. But with the strength and concentration of the drug Robin was administered, almost anyone's inner ear would be "killed," as Robin describes the action. Although injecting gentamicin directly into her inner ear was almost certain to destroy what little was left of Robin's hearing, the decision to proceed was in some ways very easy. "It was not even a choice," Robin says. "I was tired of being nonfunctional."

Over the years, falling has become routine for Robin. "There are good falls and bad falls," she says. "The ones where I've hurt myself have actually been easy falls because you're not expecting those, you're not prepared for those falls mentally. It's probably hard to understand how you can be prepared for a fall, but you can if you have done it enough." About once a week she'd fall while walking out to the parking lot from her evening college class; now a security guard escorts her to make sure she doesn't injure herself. Standing in the bathroom of her home once, she lost her balance and fell through a set of glass shower doors. Also at home, she had what she calls a "drop attack" just as she started down the stairs from the second floor. Unable to hang on to the handrail, she fell all the way to the bottom. In both these falls at home, she somehow avoided injury. She says Karen has taught her how to fall, how to break the impact with her arms. But Karen also warned her that she was playing Russian roulette; eventually one of these falls would be catastrophic.

Though it was an easy decision to move toward being more functional, the possibility of becoming stone-deaf left Robin ambivalent. She says she felt worse about the idea of not being able to hear than she did when she was told, several years ago, that she had terminal breast cancer (from which she has recovered). Listening to music (any genre except opera and rap) and playing the guitar were among her favorite activities in the world.

Robin went into the doctor's clinic on a Friday to receive the injection. She awoke Sunday morning in "a different world." She knew something wasn't right when she failed to hear her family's three cats, who are usually "as vocal as dogs" first thing in the morning, meowing at the door to be fed. "And I remember my daughter coming down from upstairs, getting ready to go to church," she says, "and I was just crying. She knew right away what was wrong."

It was a very emotional day, and for the first time in years she didn't go to church. "How was I going to hear the sermon?" she asks. But it didn't take long for her to adapt. Because her hearing had been gradually diminishing for years, Robin was at least somewhat prepared for the complete absence of sound. For one thing, she was already an expert lip reader, so it was a simple matter at church, for instance, of just sitting in the front pew, where she could see the minister's lips. She found that she could even sing along with the congregation by following the minister's lead. For her classes at the community college, she arranged for a professional transcriptionist, provided by the school, to sit next to her and take lecture notes. Although she could lip-read most of what her instructors said, often they didn't face the class when they spoke, or they played videos that she of course couldn't hear.

A few months after the first antibiotic injection, I met with Robin again. I wanted to find out how she was progressing and how her life had changed since losing her hearing. She invited me to tag along with her to a weekly bowling league session. Like most bowling alleys, Paradise Bowling Alley was a cacophony of tumbling pins, rolling balls, and excitable bowlers. As we sat on stools at a table perched a few feet above the lane to which she and her team were assigned, I was struck by how normally our conversation proceeded. Robin spoke in clear sentences, not too loud, not too soft. Her lip-reading skills were so good that I never had to repeat anything, nor did she even need to be looking directly at me to discern what I was saying; all she required was the sight of my mouth in her peripheral vision. I think I had a harder time hearing what she said, due to the noise, than she had understanding me. When it was her turn to bowl, she would stand up, pausing as she placed a hand on the low wall behind her stool, then move to the steps and grasp the handrail as she went down. After briefly touching the back of a chair, she would make her way to the trough where her ball waited. Like navigating a familiar room in the dark, this way of walking is a deliberate strategy to prevent falls, each touch stabilizing her. She learned it from Karen, who had asked Robin to log all of her tumbles, and then analyzed the patterns. Karen found that most of her falls came after quick movements, such as standing up. "So she taught me when I stood up to stand there for a moment," Robin says. "Refocus. You know how for most people you stand up and go? Not for me. For me it was stand up, get my bearings, get my center of gravity, and then move."

The movements involved in bowling—holding a heavy ball in one hand, rushing forward, and hurling it toward the pins—ought to be problematic for someone with a balance problem. Yet Robin manages surprisingly well today. She says that she has fallen while bowling but usually can "recapture" her balance with subtle movements before going down. I ask her if bowling isn't risky for someone with Ménière's, and she says it probably is but she refuses to give it up. A bowler since childhood, Robin was once so good that she toured with the army bowling team. Now it's the last of the many physical activities—from marathon running to hiking, baseball to weightlifting—she can still do, and she's not going to let the disease rob her of this pleasure too. Having Ménière's, she says, "is almost like taking away your legs because you can't do those things with your legs that you normally used to do." Balance was something she never thought about before she contracted the disease. "You don't know it's even there until it's not there," she says. "Having been a gymnast—I used to do the balance beam—I relied a lot on my balance. You had to. It's four inches wide and way off the ground. You just took balance for granted."

Robin tells me that the injection is doing its job: her vertigo attacks, which used to happen at least once a day, are down to about one or two a week. But the downside is that the attacks, unexplainably, are much stronger now than they used to be. Her doctor's goal is to reduce their occurrence to once a month. She'll need a couple more injections to kill off the few remaining functional cells in her left inner ear.

Adapting to her deafness has not been easy, however. At

Christmas, she and her two children flew back to Michigan, where she grew up. They attended a party with "the whole family, all the grandkids, everybody, forty or fifty people, and I felt like I was in the room all by myself." The hardest part has been at home, where communication can still be difficult with her family. Her husband still talks to her when he's watching television in the living room, not realizing that she can't hear him from the kitchen. "My kids are getting better about making sure they get my attention before they start talking," she says. "Every once in a while my daughter will yell down the stairs, and then she'll come down and say, 'Why didn't you answer?' And I'll just look at her."

The best news Robin relates is that she may soon be able to regain some of her hearing. She pulls back her hair to reveal the aftermath of a recent surgery. A small, shallow, rectangular section of her skull has been carved out behind her right ear, the side that isn't getting antibiotic treatments. Surgeons are in the process of attaching a specially designed sound transducer directly to her skull called a bone-anchored hearing aid, or Baha. It will allow sound to travel to her hearing nerve directly through the bone, bypassing her faulty inner ear. When she tried out the system before the surgery, using a headband-mounted version, the results were spectacular. "I heard better than I have in years," she says. "I was so ecstatic!" Once her hearing comes back, Robin won't have to rely on the transcriptionist to help her at school, and her feelings of isolation should be eased. Regardless, she'll continue with her goal of getting a two-year associate degree at Pierce College, then applying for a master's program in social work for people with hearing impairment. She thinks her experience as a hearing-impaired, and now a deaf, person will give her more empathy and compassion when working with members of this population and will lead to easier acceptance into their world.

What physical characteristics distinguish somebody with a vestibular disorder? If Irene or Robin or Cheryl or Vincent passed you on the sidewalk, could you make a diagnosis? That depends on a lot of things: the severity of the disease, how long a person has had it, the age at which it was contracted. Some people who use walkers or canes have vestibular disorders, but so do many other folks suffering from other ailments. Clues do exist, but they're usually subtle, visible only to someone who knows what to look for, like Karen Perz.

"People who are off-balance have a certain presentation," she says. "If they don't know where they are in space, then typically what you'll see is when they're walking their feet kind of drift in and out, with an uneven step width. Or their cadence will be irregular. Or their arms will be out. Or some combination of this. People who have *mal de débarquement* still feel like they're on the boat, but it doesn't look much different from someone who has an inner-ear infection."

Most people with normal balance depend on three sensory inputs to maintain an upright posture: vision, proprioception (which I'll explore in detail in the next chapter), and the vestibular system. Each one gives the brain information about the body's orientation and about the direction of earth's gravitational pull. In everyday situations, like walking on level, firm ground with good lighting, all three components "agree" with one another. How much each contributes depends on environmental cues. For example, if the walking surface changes — a path angles steeply down a mountainside or traverses a spongy peat bog — visual and vestibular inputs assume greater importance. When visibility is reduced, say at night or when exploring a cave, proprioceptive and vestibular inputs become more dominant. If the head were suddenly to accelerate, for instance by someone pushing you from behind, signals from the vestibular system would become paramount, activating postural reflexes to try to keep you from falling or at least to minimize injury if you did fall.

The brains of people with chronic vestibular dysfunction, as you would expect, rely more heavily on inputs from proprioception and vision. To remain upright, they need to have their feet planted on solid, horizontal ground, or to maintain accurate visual references at all times. This can lead to problems because the earth isn't always solid and horizontal, and visual cues can sometimes be deceiving. So one of the most important goals of vestibular physical therapy is to get the brain to compensate for the sensory input it used to rely on from the vestibular system.

This highly specialized form of physical therapy began in the early 1940s in England, when the vestibular expert Terence Cawthorne, along with another physician, F. S. Cooksey, established a protocol for the rehabilitation of patients with vestibular injuries and head traumas. "The symptom complex for which [we] had to provide," Cooksey wrote in 1945, "consists of headache, vertigo including the so-called 'black-outs,' impaired mental concentration, and deafness in a proportion of cases. Of these symptoms the vertigo may be the most disturbing; but, fortunately is usually amenable to treatment." He went on to state that these symptoms often lingered for much longer than one would suspect, sometimes for a year or more. This delay was due, he believed, to a failure to recognize the vestibular origin of the symptoms, together with a "failure to provide adequate measures to restore confidence" in the patients, which was often severely damaged as well.

Cooksey and Cawthorne devised a series of physical exercises that, with a few exceptions, are nearly identical to those in use today. His explanation of their purpose also sounds similar to today's approach. "The exercises are designed to restore balance as far as possible and to train the eyes and muscle and joint sense [proprioception] to compensate for permanent vestibular dysfunction. Because so many patients are worse in the dark we pay special attention to muscle and joint sense by performing many exercises with the eyes closed." He described a series of movements that began when patients were seated or in bed. They started with simple eye and head motions, graduating to shoulder shrugging and leaning forward and picking up things from the floor. Those who could stand were directed to walk up and down steps and ladders, with eyes open and shut. Next came "games with balls and bean bags, when the instructor tosses a ball, high or low, which the patient catches, holds above his head, turns smartly about and bends down to throw the ball back to the instructor between the legs." Then, usually in a group setting, the patients would form a circle around someone, walking slowly, catching and returning a large ball. This regimen became known as Cawthorne-Cooksey exercises.

The exercises were effective then, and remain so, because whenever any of the body's three sensory inputs for balance becomes dysfunctional, the brain can learn to adapt by relying more on the remaining intact senses. If, say, proprioception in the feet is diminished, perhaps due to diabetes or aging, then the brain can adjust by focusing more on vision and vestibular inputs. But this process is akin to the way muscles get stronger: by being challenged. And though the brain can adapt, thanks to its innate "plasticity," or ability to restructure its connections, most people are loath to put themselves in a precarious situation. For in order to challenge the balance system, it's usually necessary to expose oneself to a stimulus that creates dizziness or imbalance, which can be extraordinarily uncomfortable.

Though some old-school physicians are skeptical, evidence is mounting that this approach not only is effective but works on people with exceptionally difficult problems. According to Lucy Yardley and Linda Luxon, two British vestibular researchers, clinical trials have shown improvements of symptoms in over 80 percent of the people who undergo vestibular physical therapy. Though it's not a cure, about a third report complete elimination of vertigo and dizziness after treatment, the authors stated in an editorial in the *British Medical Journal* in 1994. That accords with data accumulated in studies conducted over the past few years cited by Fay Horak, a neuroscientist and physical therapist at the Neurological Science Institute, in Beaverton, Oregon. One of the leading balance experts in the country, Horak said recent research demonstrates that balance physical therapy can benefit even people in their nineties, or who have had a stroke, or who suffer from Parkinson's disease. Not only does it improve balance in daily activities, but, perhaps more importantly, it reduces the risk of falling.

When a patient with a balance problem comes in to see a vestibular physical therapist like Karen Perz, the first thing she'll do is conduct an interview to find out when and how the imbalance manifests itself. Then she'll do a detailed analysis of a patient's baseline balance, using a battery of simple tests, such as how long a person can stand on one leg, with eyes open and closed. If necessary, she'll proceed to a more detailed investigation, which may involve diagnostic tools such as the NeuroCom Balance Master. Resembling an oversized telephone booth with the door removed, it's a highly sophisticated machine used to measure the balance of returning NASA astronauts. A computer controls the movement of both the platform and the visual field, so that when a person is harnessed in (to prevent falls), the device can accurately measure the contributions of each of the three sensory inputs for balance.

One of the patients Karen tests on the Balance Master today is a woman I'll call Kate. She is a medical transcriptionist, in her fifties, with an otherwise clean bill of health. A few years ago she began having vertigo attacks after painting a ceiling in her home. Since then, she's had two episodes in which, she says, suddenly "my arms and legs didn't even belong to me. I said, Lord, I know you don't mean for me to lay on the floor...It's like somebody threw me into a swimming pool in the dark; you don't know up from down." After talking to her for several minutes, then observing her as she walked across the room and up and down a short set of stairs, Karen had Kate step into the NeuroCom machine, buckling her into the safety harness. The computer automatically ran a series of diagnostic tests, moving the platform in different ways, as though an earthquake were shaking it. First the solid walls, painted with a simple representation of mountains, lurched back and forth, then the platform rumbled, then both shook together. The test took only about ten minutes. What it revealed confirmed Karen Perz's suspicion: Kate's balance system relied too much on her vision.

The therapy that Karen outlines for Kate, which will take place over the next four to eight weeks, involves training her brain to downplay visual inputs, so that, for instance, she can look around as she walks and not be thrown off by moving objects. To perfect this skill, one unusual technique Karen employs is to have Kate look at a large sheet of checkered paper on the wall. On it is pinned a piece of paper with a word written on it. Kate is instructed to look at the word while moving her head back and forth. After a minute, Karen asks her if this exercise is making her dizzy, and Kate nods. "What this does," Karen explains, "is confuse the visual system. It's a way of forcing your brain to pull in your vestibular system a little bit more because the visual system is being deliberately confused. This is helping your brain to increase its awareness that vision is not always a good reference point." Karen asks Kate to purchase a standard red-and-white-checkered picnic cloth and hang it on a wall at home so she can practice the exercises on her own. "Checkers are probably the best," she says, "because they're more confusing to the visual system. Anything with a really strong pattern will work too. A strong plaid will work."

This technique of having patients move their head back and forth while walking or standing seems to be a common denominator in several of the cases I watch Karen handle. I asked her later when she applies it and how it works. Normal people, when looking at something while moving their head back and forth, are able to keep their eyes focused on the object because of the vestibular ocular reflex (VOR). "But if there's a weakness in the vestibular system," Karen elaborated, "then when you turn your head quickly your vision will not stay stable because this reflex is weak. The image does not stay stable on the retina. The brain recognizes this defect and begins to boost or amplify the signal from the weak side in order to stabilize the system. The advantage to this exercise is that you've eliminated almost all other variables. With standing on one foot, there's all kinds of variables. But the ability to stabilize your vision during head motion — it doesn't go through very many connections. By giving the brain repetitive information that 'Hey, my eyes aren't staying steady during movement,' the brain will adjust." Kate's dizziness and vertigo attacks stem in part, Karen says, from her brain's inability to integrate the sensory information. For some unknown reason, Kate's brain gives too much priority to her visual input, which in certain circumstances can be easily fooled, even in people with no dysfunction.

The classic example of this deception occurs when you're in a car stopped at an intersection. If you're not paying attention and the car next to you moves forward slightly, your brain may think that *your* car is rolling *backward*. In "normal" people, the vestibular system would then kick in and inform the brain that in fact no motion has been felt. Kate experiences this type of illusion frequently, often in grocery stores and Home Depots, where people are moving past her. Her brain interprets the sight of this movement as her own motion in the opposite direction, and she can become disoriented and unstable. "If you're dependent on vision for your sense of balance," Karen explains, "it's hard to be in an environment that's not visually stable. And our world is not visually stable. Cars move on the street, people walk around."

Kate has had just about every medical examination known to science, yet nobody has come up with a cause for her disequilibrium. Her doctor basically said he just hoped she got better. Yet these attacks are as disturbing to her as they are to someone diagnosed with Ménière's disease or bilateral vestibular disorder or positional vertigo. She is in constant dread because she doesn't know when the next episode will come. "I think about it all the time," Kate says. "I'm concentrating on my walk all the time. If I'm at the mall and I mindlessly just start thinking about something else, I'm going to cross my legs in front of me." About once a day she has to reach out and grab something to hang on to so she won't fall, yet she's fallen only once in the past three months. Brief interludes of stability are like a gift: "There are times when suddenly it's like the door of the dungeon opens and I can run out, and I feel great. I no longer feel that gyroscope in my head, that constant spinning, like something in my head is just so overexcited."

Kate's disorder doesn't seem to affect her performance on the job, though she's thankful she works at home and not at the health care facility that employs her. As she's transcribing doctor's notations at her computer, sometimes her vision gets blurry and she becomes exhausted. "If I had to work at the office," she tells Karen, "I would have been fired. But because I'm home I can get up, reorient myself. It gives me freedom if I have to lay on the floor for a few minutes or do a little mental relaxation. If I was in the office, they'd be saying, 'What's Kate getting up every half an hour for?'"

Kate's experience sheds light on just how complicated the balance system is and how difficult it is to treat a balance problem. When her doctor pronounced that her neurology tests had checked out fine and there was nothing he could do for her, he didn't sound much different from van Gogh's doctors after they examined him, or Robin Grindstaff's army doctors after her initial diagnosis of Ménière's. When Kate decided to pursue vestibular physical therapy, she was in something of a crisis mode. Nobody in the traditional medical world had been able to help her. The only relief she had gotten, and that only temporary, had been from a chiropractor who had adjusted her cervical spine, believing that her balance problems stemmed from misaligned vertebrae. But many traditional doctors are skeptical of such treatment, at least until more studies have been done that support its effectiveness.

What Karen Perz offers in her clinic is a way to help retrain the brain to adapt to a different regime of sensory input. No simple deduction led to the treatments she uses today. Rather, the incredible complexity of a balance system that is designed to coordinate three separate inputs stymied scientists until very recently. Let's backtrack a few centuries to see how we've come to learn about the balance system, and why it took so long to give up its secrets.

Chapter Three

The Spin Doctors and the Discovery of "Multimodality"

If Aristotle failed to take notice of balance as a sense, then Galen, the Greek anatomist and physician, would at least begin to examine some of its basic physiology five hundred years later. While dissecting dogs, pigs, and apes during the second century AD, he began to unravel many mysteries of mammalian biology, including the existence of the inner ear. Its strange curving structure reminded him of the Labyrinthos, a mazelike edifice built, according to myth, beneath the palace of King Minos of Crete.[1] It was designed to house the ferocious Minotaur, a man-eating monster with the head and tail of a bull and the body of a man. As many explorers of a new territory are apt to do, Galen assumed the right to name what he discovered, and the inner ear is still called the labyrinth today. But Galen, like scientists and physicians for the next 1,600 years, could surmise only that the labyrinth might have something to do with hearing. Its role as an organ of equilibrium wouldn't be unveiled until the nineteenth century.

One of the first "modern" men of science to investigate the balance sense was Erasmus Darwin, Charles Darwin's grandfather. He studied the effects of vertigo and nystagmus caused by rotating the body. In performing his experiments, Darwin, described by historians as stoop-shouldered, corpulent, and fond of wearing a haphazardly placed wig, used himself as a subject, spinning around and around to induce dizziness.[2] When he published *Zoonomia; or, the Laws of Organic Life* between 1794 and 1796, Darwin devoted an entire chapter to the subject of vertigo, though some scholars believe he did more to confuse than to illuminate the issue. He talked about three types of vertigo: "flight" vertigo, he wrote, was caused by indigestion; "irritative muscular motions, such as those of the stomach from intoxication," were reported to be the root of a second type of vertigo; and auditory vertigo, evidenced by a "noise in the head," was the result of diminished hearing abilities.[3] During his research, Darwin, presumably to make the task of rotating easier and more regulated, hit upon the notion of a spinning machine. He even had drawings made of such a device, but he never went further.[4]

In the late eighteenth century, however, a British doctor named Joseph Mason Cox, said to be the first physician to specialize in treating the mentally ill, heard of Darwin's idea. Around 1788, he decided to build a spinning machine to use on patients at Fishponds Private Lunatic Asylum near Bristol, a facility owned by his family and managed by Cox himself. Soon after he began using rotation as "both a moral and medical mean in the treatment of maniacs"[5] and wrote one of the first books on treating mental illness, hospitals and asylums across England and Europe began using the therapy, which remained popular until the mid-nineteenth century. Commenting on the utility of the treatment, Cox wrote that it was a "mechanical anodyne. After a few circumvolutions,

I have witnessed the soothing lulling effects, when the mind has become tranquilized, and the body quiescent; a degree of vertigo has often followed, and this been succeeded by the most refreshing slumbers; an object this the most desirable in every case of madness, and with the utmost difficulty procured."[6]

Although today we would think of such treatment as inhumane, in the early nineteenth century it was viewed as something of a godsend, considered superior even to opiates for calming patients. It didn't matter that no one really understood how the therapy worked. Because vomiting was considered one of the "most successful remedies in madness," Cox wrote, "if the swing produced only this effect, its properties would be valuable; but though it can be employed so as to occasion the mildest and most gentle effects, yet its action can be so regulated as to excite the most violent convulsions of the stomach." Apparently, the stronger the effect of the "anodyne," the more remedial it was considered to be. Some asylums later built spinning machines that could handle four patients at a time, rotating them at up to one hundred revolutions per minute. Although there's no record of how the patients themselves felt about the procedure, a comment from Cox hints at their general reaction: "The impression made on the mind by the recollection of its action on the body is another very important property of the swing, and the physician will often only have to threaten its employment to secure compliance with his wishes."

When nineteenth-century mental patients were spun in a rotating chair, illness was induced, as it is in motion sickness, because a conflict arose between two of the three sensory components of balance: the eyes reported to the brain that the body was moving, while the vestibular system indicated that the body was still. If patients had only learned to close their eyes during the spinning treatment, they would have spared themselves untold agony. But the sensory conflict mechanism wasn't known to physiologists in the nineteenth century. The vestibular system was still thought to have no function other than in hearing. The body, scientists believed, sensed its own movements either by the distribution of blood, through pressure receptors in the skin, or directly through the brain—most likely in the cerebellum, because this area was thought to coordinate physical movement (experiments in the mid-nineteenth century demonstrated that when the cerebellum was removed, animals had trouble with this function).

It was not until several research scientists in the 1860s enlisted the aid of rotary devices, like those used in asylums and hospitals, that the mysteries of the vestibular system began to be revealed.

One of the first to use spinning machines for pure research was the Czech physiologist Jan Evangelista Purkyne.[7] During his studies of vertigo and eye movement, which took place in about 1820, he even made use of a children's *Ringelspiel,* or carousel, at a Prague park. His rotational experiments, both on himself and on mentally ill patients, showed that the head, and not some other part of the body, was the source of vertigo. He pinpointed several different causes of the condition, among them alcohol, electricity, heights, motion, and visual stimuli alone, without bodily motion. Purkyne's work firmly established the relationship between vertigo and the reflex movements of the eye (nystagmus). By altering the position of the head during and after rotation, he could observe a change in the direction of these eye movements. This phenomenon hinted at the action of individual semicircular canals; each would fire most strongly when the rotation took place in the axis it occupied, then activate a corresponding set of eye muscles. But because the function of the canals wasn't yet known, the disparate eye movements were regarded as an unexplainable curiosity.

Until this time, no special sense organ for equilibrium had been discovered, although a French physiologist working at the same time as Purkyne came tantalyzingly close. His name was Marie-Jean-Pierre Flourens. In Paris, while investigating how various parts of the brain functioned, he pioneered the technique of "lesioning." He would selectively destroy a certain area of the brain and then make observations about the effects.[8] Lesioning the semicircular canals of a pigeon produced some startling responses. Though Flourens had predicted that the bird would be unable to hear after the surgery, instead it walked in bizarre patterns, occasionally lost its balance, and hid in dark corners. When the horizontal semicircular canal was destroyed, the bird walked in endless circles. Destroying the vertical canal caused the bird to turn somersaults. It could drink only by holding its head upside down. This experiment established for the first time that the canals weren't involved in hearing. Amazingly, Flourens didn't realize that the birds' unusual behavior was caused by dizziness, perhaps because the symptoms of vertigo in birds were different from those of humans. So the cerebellum, rather than the vestibular apparatus, remained the imagined source of vertigo.

If Flourens and Purkyne had known of each other's research, they might have collaborated and discovered the true function of the vestibular system. As it was, progress in that direction would have to wait another thirty years, until Prosper Ménière, another Frenchman, made a small but important contribution. Ménière was an otologist working at the Imperial Institution for Deaf Mutes in Paris. There he saw patients in whom the hearing part of the vestibular apparatus, the cochlea, had been impaired, leading to symptoms of tinnitus and vertigo. In 1861, recalling Flourens's experiments on pigeons, Ménière made the bold claim before the French Academy of Medicine that the inner ear was the seat of vertigo. Although his hunch proved correct, he died that year, before anyone could verify it.[9]

Another nine years passed before a German physiologist, Friedrich Goltz, became the first to assess the true function of the canals, after repeating Flourens's experiments on pigeons. Goltz reasoned that if destroying the canals produced vertigo and a lack of balance, then the organ's normal function must be to regulate equilibrium. Goltz was right, but his ideas about how the canals worked were flawed.[10]

That final problem of how the canals functioned was left to three men, working independently, to solve. Like separate paths that suddenly merge in the forest, these three—a physicist, a physician, and a chemist; two in Austria and one in Scotland—published nearly identical papers at nearly the same time, between 1873 and 1874.

The first was Ernst Mach, a native of Bohemia, who began his career as an apprentice cabinetmaker but, failing at that, went on to study physics and math at Vienna University, earning a doctorate in experimental physics.[11] Among the subjects of his many wide-ranging investigations was supersonic velocity. Indeed his name is known to many people today through the Mach number, which is the ratio of the speed of a projectile, such as a jet plane, to the speed of sound, as in Mach 1 or Mach 2. Albert Einstein credited Mach with paving the way for his theory of relativity, and he said in an obituary at Mach's death in 1916 that "even...those who think of themselves as being his opponents hardly know how much of Mach's scientific approach they, so to speak, sucked in with their mother's milk."[12]

For a small part of his illustrious career, Mach studied the physiology of sensation and movement, including a thorough look into the mechanism of the semicircular canals. After working out the mathematical equations for rotary movements, he looked for evidence of a human sensory organ capable of perceiving rotary motion. He refused to believe, as Purkyne had, that the brain directly perceived rotary forces acting upon it. One by one, Mach eliminated each sense as being capable of detecting such turns. Only one remained: the semicircular canal apparatus. He went on to show that, because of its peculiar structure, the semicircular canals were adapted to the task of perceiving rotary movement.

Mach arrived at his computations by doing experiments in fluid dynamics. He carved into a small brass plate an impression the size of a semicircular canal, filled it with fluid, added small particles to the fluid so that he could detect its movement, and covered the device with glass. Then, using a small centrifuge, he spun the model around and observed how the fluid shifted. From this device, he estimated the energy required to move fluid during rotations. One of his major findings was that humans are able to sense only certain kinds of motion. We can detect acceleration and deceleration, but not constant velocity. A familiar example occurs when you ride an escalator. Stepping onto the machine, your vestibular system tells your brain that the body is accelerating forward. But once you're moving at a steady pace, the vestibular system falsely indicates zero motion. If you were to close your eyes, you would think you were standing still. (This physiological quirk has fascinating consequences for air travel, as we'll see later.)

In Vienna, before a meeting of the Academy of Sciences in 1873, Mach announced his discovery. But then a strange thing happened. Also attending the conference was a man named Josef Breuer, a Viennese physician and physiologist, who, after hearing Mach's report, asked that he be permitted to publish a brief report of his own studies. The following week, Academy members were mildly shocked to learn that Breuer had been researching the same subject as Mach and had reached nearly the same conclusions.[13] In his research on the vestibular apparatus, Breuer had duplicated Flourens's experiments, lesioning the canals of pigeons. His improved technique, however, led him to the idea that it was movement of the endolymph, or fluid within each canal, that caused the strange effects on the birds' mobility. In addition, he had designed a method of rotating animals in a miniature version of Cox's Fishponds Asylum spinning machine, which caused the same effects in animals as lesioning their canals, with the benefit that it was temporary and harmless.

Mach and Breuer had arrived at a common understanding of how the semicircular canals and otolith organs worked. The two men firmly proved that these unusual body parts were singularly dedicated to maintaining equilibrium and weren't at all involved in hearing. Throughout the scientific community, this idea became known as the Mach-Breuer hypothesis. This coinage was less of a mouthful to pronounce, perhaps, than Mach–Breuer–Crum Brown Hypothesis, which would have paid homage to the third scientist to reach the same view of how the vestibular apparatus worked.

Alexander Crum Brown was a thirty-six-year-old Scot with a master's degree in chemistry and a Ph.D. in medicine, though he never practiced the latter. A brilliant though somewhat disorganized professor at the University of Edinburgh, Crum Brown, with his lively black eyes and trademark skullcap, left most of his students gasping for intellectual breath, as one former student remembered: "Briskly entering the class-room, he began at once in rapid phrasing to describe the properties of a chemical substance or the intricacies of a chemical process. Chemical formulae grew like magic on the black-board. The casual and limp-minded listener found Crum Brown's quick vivid style much too strenuous; but the student who really wished to learn, and had ear and eye in well-trained attention, could not fail to experience keen intellectual delight from the masterly manner in which the whole subject was presented."[14] Perhaps because most of his students fell under the "limp-minded" category, his classes were notoriously boisterous and undisciplined. As another student said, "Crum Brown was a charming man and a very bad teacher. Most of his students very soon gave up all attempt to follow him and the class was exceedingly rowdy. Some days the noise and interruptions were so great that the poor professor had to give up and flee. Then in a few minutes he would return with tears streaming down his cheeks and apologise for his inability to control his class."

It isn't clear why Crum Brown, who made his mark by inventing symbols used in writing chemical formulas, studied the physiology of semicircular canals. Like the other two men, he attacked the problem by rotating his subjects on a specially built table and then analyzing the effects. Unlike the others, he used only humans in his experiments, blindfolding them first and placing their heads in a variety of positions. From the data he gathered, which he published in 1874, Crum Brown essentially corroborated the ideas generated by Mach and Breuer.

So was it just a coincidence that these three men, from different corners of the scientific universe, only one of whom was a physiologist, would come up with the key to the function of the vestibular system at the same time? There had been a steady progression of knowledge from Purkyne to Flourens to Ménière to Goltz, but nobody had been capable of completing the puzzle. These earlier pioneers of the vestibular system, except for Ménière, were all physiologists.

Perhaps the problem required a different perspective. Mach, in 1875, intimated as much when he wondered how a physicist like himself—or, for that matter, a chemist like Crum Brown—could make contributions in physiology. "The facts which led to my theory and which are sufficient to explain it have all been known since 1824, that is more than half a century. If such an interpretation has not been made yet, then this is proof that physiologists are often not familiar enough with even simple physical considerations. Therefore I feel entitled to have a word in these matters as a physicist. Although the physiologists might not agree with all parts of my work, consideration of the physical side of this subject should not prove to be completely without value."[15]

About a decade after Mach and his contemporaries published their studies, William James presented his paper "The Sense of Dizziness in Deaf Mutes." While other scientists were busy showing that the vestibular apparatus had nothing to do with hearing, James upset the cart by demonstrating that there was indeed a connection. His experiments proved that a high percentage of deaf mutes had trouble with balance, though he couldn't explain why. It was not until many years later, when technology permitted a more microscopic examination of the inner ear, that scientists observed that defects in the hearing part of the inner ear (the cochlea) might also be shared by the balance part (semicircular canals and otolith organs), both of which share a common connection to the brain (eighth cranial nerve), which could be faulty.

At about this time, an English physiologist who would later win a Nobel Prize in Medicine, Charles Scott Sherrington, was making pivotal discoveries about the body's reflex system that would contribute to the understanding of balance. In his classic work, *The Integrative Action of the Nervous System,* published in 1906, Sherrington unveiled a new concept. He told of a new type of sense receptor, cells in the joints and along-side muscles, whose sole purpose was to detect movement of the limbs. Many scientists doubted Sherrington's claims at first.[16] The sensory nerves that Sherrington described had never been found before, and it was argued that nerves on the skin's surface were adequate for relaying information about the muscles to the brain. But Sherrington, working with cats and monkeys, found a curious thing when he severed the motor nerves of their spinal cords. A large percentage—up to one half—of nerves originating from limb muscles didn't stop functioning or degrade. What they were doing, Sherrington reasoned, was registering positional information about the muscles and sending it to the spinal cord.

Sherrington called this system by which the body senses its own motion *proprioception*. (*Proprio* is Latin for "one's self.") It can be thought of as a variation on the sense of touch. Proprioceptive nerves detect the movement of muscles and joints at a subconscious level. For instance, close your eyes and move your hand around in front of you. At all times you know exactly where your hand is without the involve-ment of any of the five traditional senses (unless of course you're wearing heavily perfumed hand cream or are snapping your fingers at the same time). He believed that this system, which some contemporary scientists hailed as a "sixth sense," contributed to the body's reflex system for maintaining an up-right posture. Sherrington's discovery added proprioception to vision and the vestibular system to form the triumvirate of sensory inputs for balance.

Robert Bárány, an Austrian otologist, also made large con-tributions to the field of balance research during the early part of the twentieth century. One was a method for testing the function of the semicircular canals.[17] While working in an otology clinic in Vienna, Bárány routinely performed a procedure called syringing, or irrigating a patient's ears with liquids. He noticed that when he did this some patients complained of vertigo, as evidenced by their nystagmus, the rapid eye movements that reflexively occur when the vestibular apparatus is overly stimulated. "I then realized that some general principle must be implied, but at the time I did not understand it," he said. Then a patient accidentally revealed the answer to him. While Bárány was syringing his ear with cool water, the patient said, "Doctor, I only get giddy [dizzy] when the water is not warm enough. When I do my own ears at home and use warm enough water, I never get giddy." So Bárány ordered the nurse to bring him a bowl of warmer water, and when he began irrigating the patient's ear, the man protested: "But, Doctor, this water is much too hot and now I am giddy again." He quickly observed the patient's eyes and saw that they were in nystagmus, but in the direction opposite to that caused by the cold water. "It came to me then in a flash," Bárány said, "that obviously the temperature of the water was responsible for the nystagmus."

Why should temperature matter? Bárány was at a loss to understand until he recalled a physical principle he'd first observed in childhood. "I remembered the bath water tank and my surprise, as a child, at finding the water immediately above the fire quite cold, whereas higher up, at the top, the tank was so hot that it burned one's finger. The labyrinth reminded me of a bath water tank, i.e. a container filled with fluid." He imagined that by irrigating the ear with liquid cooler than body temperature, the inner-ear fluid, or endolymph, would move in a certain direction, triggering nystagmus. When warmer liquid was used, the endolymph would flow in the opposite direction. The nystagmus, too, would reverse direction. If correct, this mechanism would confirm the ideas of Mach and Breuer about the role of endolymph movement in stimulating the canals. After performing a series of tests, Bárány validated his hypothesis, and the caloric reaction test was born. It remains in use today as one of the easiest ways to assess the condition of the semicircular canals.

Also still employed frequently today in many vestibular clinics is a more technically sophisticated version of the Bárány Chair, a rotating device not much different from the ones used by nineteenth-century researchers, but which Bárány improved and standardized. Drawings of the chair, such as the one made by the Physician's Supply Company of Philadelphia, show a simple unpadded chair with arms and low back, a footrest in front, a pedal to raise or lower the height, and a bar across the front to keep the occupant from falling out.[18] In back is an adjustable and quite uncomfortable-looking head brace, and beside it a vertical pole with which the physician can twirl the chair about its axis. Today the chair is used by clinicians to test the health of the semicircular canals. After being spun at a constant rate in one direction for about a minute, the chair is then suddenly stopped. The patient's ensuing nystagmus indicates whether the canals are working properly.

Mainly for these two advances, Bárány was awarded the 1914 Nobel Prize in Medicine, but when the awards were announced the young Austrian couldn't at first be found. A soldier in the German army, he had been captured by the Russians during World War I and interned as a prisoner of war. Only through the intervention of Swedish diplomats was he freed from the prison camp and allowed to attend, two years after the award was announced, the Nobel ceremony in Sweden, where he ended up spending most of the rest of his life.

Despite all the advances in understanding the vestibular system and proprioception that occurred in the nineteenth century, much about the human balance system was still veiled in mystery during the first part of the twentieth century. One of the leading American otologists of the time, Dr. Isaac Jones, reported that in the early 1900s, medical schools barely touched on the vestibular system: "Those of us actually studying the ear in medical school were taught, 'The ear also has semicircular canals. They are concerned with equilibrium—we will now take up the study of the heart.' That was about all they had to tell us."[19]

Perhaps this dearth of solid information led to some of the more infamous misconceptions about the inner ear, such as the one Jones himself perpetuated when he claimed that the vestibular system was absolutely essential for aviators to fly. In his book *Equilibrium and Vertigo*, published in 1918 and adopted as a standard text for the medical division of the Army Air Corps, Jones and his coauthor, Lewis Fisher, wrote:

When flying through the air, on what does the aviator rely in order to maintain his equilibrium and that of the aeroplane? Can he rely on sight? Hardly, for when he is sailing through the clouds or darkness, his eyes cannot give him the slightest information about his position in space—not even whether he is right side up or upside down. As regards the muscle-sense [proprioception] it is undoubtedly true that it plays a certain part; but when the aviator is seated on an unstable and rapidly moving machine, it is hardly conceivable that the weight of his body could determine and maintain his position in space merely by the sensing of gravity. In order, therefore, to preserve that wonderful accuracy necessary in controlling such a delicate mechanism as the flying machine, he relies pre-eminently on his ear balance-sense.[20]

Drawing a comparison between birds and what he called "bird-man," meaning a pilot, Jones made a spectacular error when he stated that birds rely on their "wonderfully well-developed" semicircular canals to enable them to fly through clouds. "[A bird's] muscle-sense naturally means nothing to him; his sight is of no help. He relies exclusively, therefore, upon normal functionating semicircular canals."[21] While Jones's tone is authoritative, his facts were wrong. Army pilots were later involved in experiments to test this hypothesis. They flew into clouds, tossed pigeons overboard, and watched what happened. Every bird went into a sort of emergency mode, setting its wings for an autopilot glide to the ground. "The homing pigeon, born to fly, refuses to fly if he runs into a fog and cannot see," said a 1944 U.S. Army Air Force publication for pilots entitled *Your Body in Flight*. "Many a good man has died trying to prove he was better than the homing pigeon."

Fortunately, Jones's misconception was erased a few years after he proclaimed it, when it became clear to him (and other flight surgeons, presumably) that vision was the most important sense for maintaining equilibrium in the air. Without apologizing or even owning up to his previous false statements, Jones, in 1937, wrote: "Any over-confident pilot who still feels that he can rely upon himself and his own senses in meeting all the conditions in the air is in danger. He is apt to come to grief. Through all the ages the bird has been meeting the conditions in the air.... Yet we have found that a bird always avoids a fog or a cloud. We do not find a bird flying in the clear air above the clouds — no doubt because he did not wish to fly up through them. In short, even the bird is not at his ease when he is flying blind."[22]

Although it was known at the time that there were three sensory components of balance — visual, proprioceptive, and vestibular—it wasn't clear how their inputs were integrated and interpreted by the brain. Were the contributions of the components always equal? If not, under what conditions would one input have precedence over the others? Ironically, Jones was involved in shedding light on this question soon after World War I ended. He and two "nerve specialists," Drs. Samuel Ingham and I. Leon Myers, set up a series of experiments to determine how a falling cat is able to land on its feet. Using new technology developed for the motion picture industry, the researchers photographed cats dropped from an upside-down position onto soft mats. Then they observed the animals' so-called air righting reflexes, the ability in midair to orient their feet toward the earth. Since these reflex movements were too rapid to see with the naked eye, the researchers employed slow-motion movie cameras. All the felines in the test, when dropped from a height of several feet, were able to spin themselves around in midair and land on their feet, as all healthy cats can. Next the scientists placed hoods over the heads of cats and kittens before dropping them. Again the animals landed on their feet every time. Then, as Flourens had first done with pigeons nearly a century before, the scientists lesioned (surgically destroyed) the cats' vestibular systems. After healing, these animals were again dropped, and the film footage revealed that they tumbled "over and over until earth [was] struck."[23]

With or without sight, the lesioned cats were not able to land on their feet. Cats with only one of their vestibular organs lesioned performed slightly better than cats with both organs destroyed, but not as well as the healthy felines. Studies like this one paved the way for understanding the vestibular system's role in maintaining equilibrium through postural reflexes.

In a 1988 article in the journal *Nature,* Jared Diamond, the well-known author of *Guns, Germs, and Steel* and a UCLA professor of physiology, discussed in great detail why cats can tolerate falls from heights that would kill most humans, a subject that he says "has not received the scientific attention that it deserves."[24] He cited a study written by two veterinarians from New York City, a place where cats apparently fall with some regularity. Looking at a database of 132 cats who had fallen two or more stories (a story is 15 feet), the vets found that the maximum fall was 32 stories, with a mean of 5.5. Astonishingly, 90 percent of the cats survived. Only 11 died as a result of their injuries. The most amazing finding was that the fatality rate actually *decreased* above 7 stories. In fact, the record-holding feline, who fell from 32 stories (480 feet) onto concrete "was released after 2 days of observation in the hospital, having suffered nothing worse than a chipped tooth and mild pneumothorax." (While nowhere near as "fallproof" as cats, humans don't do as badly as one would think. In falls of fewer than 4 stories, humans have at least a 50 percent chance of survival. And most land on their feet, as cats do, though more by accident than because of the vestibular system. The second most common landing point is the head, which is also where nearly all falling human babies land, due to its great weight relative to the rest of the body.)

Why cats are so resilient in surviving falls, compared to humans, has to do with several things, Diamond says. First, they have a much smaller body mass than humans, and weigh considerably less, so their terminal velocity (the maximum speed they reach in a free fall) is about half that of humans (60 mph versus 120). Cats also reach this velocity after only about 100 feet; it is thought that once it attains this speed, a cat "may relax and extend the limbs more horizontally in flying squirrel fashion, thus not only reducing the velocity of fall but also absorbing the impact over a greater area of its body."

This reaction, according to Diamond, perhaps explains why the injury and mortality rate of cats decreases above 100 feet.

Modern humans, of course, don't have a cat's ability to right themselves in the air this way. Rather than helping us orient our bodies in midair falls, the vestibular system is important for maintaining stability, both of our posture as we ambulate across the planet and of our visual gaze. Cheryl Schiltz's experience is an illustration. When she woke the morning after the antibiotic gentamicin had effectively "lesioned" both of her vestibular organs, she collapsed to the floor as soon as she tried to get up from bed and literally had to crawl downstairs. Her postural reflexes were temporarily stunned by the sudden loss of sensory input from her vestibular system, and her vision became wobbly as well.

Weeks later, Cheryl's brain was able to adapt to the lost input from her vestibular system by focusing on the input from her proprioception and her vision. This unusual multi-modal nature of balance, whereby three sensory inputs are coordinated and integrated by the central nervous system, and the extraordinary ability of the brain to compensate for weakness or dysfunction in any one input by increasing its reliance on the other two, are probably the two greatest discoveries about the human balance system that occurred in the twentieth century.

Maintaining equilibrium requires three sensory inputs because the act of balancing a mass as large as a human body over a base as small as two human feet is exceptionally demanding. It's roughly equivalent to trying to balance a triangular object on one of its points; the natural tendency is for gravity to push it over. None of the three sensory inputs alone can accomplish this feat under all circumstances. In certain conditions each can be "deceived" into sending to the brain what are termed "ambiguous" signals. To resolve the ambiguity, corroboration is required from one or both of the other sensory inputs.

Karen Perz's patient Kate, you'll remember, was vulnerable to dizzy sensations when shopping in places like Home Depot, where the environment is full of unusual visual motion. She was prone to a type of illusion called vection, the perception of self-motion induced by visual cues. As I described before, this phenomenon can also occur when your car is stopped, and the car next to you rolls forward slightly. You may mistakenly sense that your vehicle has rolled backward and instinctively step on the brakes. If you had closed your eyes, of course, you'd sense no illusion, for the vestibular system and proprioception would have correctly informed your brain that there was no motion. Why they don't in this case has to do with your experience and expectations; because you know it's possible for a car to roll backward when stopped, your brain overrides the correct vestibular and proprioceptive inputs and instead believes the false visual input. This phenomenon would seem to mean that vision, of the three sensory inputs for balance, is the one the brain relies on as the most "trustworthy," at least in certain conditions. In this particular case, because your brain holds powerful memories of times when your car has rolled backward when stopped, the combination of memory and vision is enough to override the other two inputs.

Vection can also occur while you're watching an IMAX movie in which a camera is mounted to an airplane. You're seated in a chair with your feet firmly on the ground, yet when the plane banks to the left, say, as it soars across the

Grand Canyon, you catch yourself leaning to the left. Your eye perceives that you're "inside" the plane. After a while you may perhaps even start to feel a little queasy, the first sign of motion sickness. Those sensations are coming from visual input alone, because if you close your eyes you no longer feel the motion and thus don't feel sick. Again, it's almost as if the brain "wants" to believe what it is seeing, despite the conflicting but correct inputs from the vestibular system, which indicate that you're not moving. We humans are a highly visual species, and the sense of vision exerts an enormous influence on the contours of the brain's map of the world. But sometimes vision isn't a reliable sense for maintaining balance. It gets dark. We can't see. Some of us are blind. Yet we can still balance on two legs, thanks to the input from proprioception and the vestibular system.

Ambiguous signals also can come from the vestibular system itself. One involves something Albert Einstein discovered in 1911: because gravity is a kind of linear acceleration (which on Earth "pulls" an object toward the center of the planet at 9.8 meters per second), it is indistinguishable from an equivalent acceleration due to motion. While this idea may at first sound confusing, it's not too difficult to understand if you imagine holding a serving tray containing a dozen billiard balls. If you quickly move the tray to the left, the balls will tend to lag behind, due to inertia, and roll in the opposite direction (to the right). Now if you tilt the tray slightly to the right, gravity will move the balls in that direction, and they'll also exert a force to the right. So linear acceleration to the left produces the same effect on the balls as tilt (gravitational pull) to the right. This is roughly what happens to the vestibular system's otolith organs, the utricle and saccule, which detect straight-line acceleration of the head, one in the vertical plane (saccule), the other in the horizontal plane (utricle). Using an ingenious and very old mechanism (we share it with lobsters!), these organs consist of a layer of calcium carbonate crystals (the otoliths or "ear rocks") floating atop a layer of extremely sensitive hair cells, called the macula. In the analogy I've used, the billiard balls represent the calcium carbonate crystals, while the tray signifies the macula. The otoliths by themselves can't tell the difference between a linear acceleration (as when you're merging quickly onto the freeway in a car) and a head tilt. The only way to resolve the ambiguity is with other senses. Your eyes quickly tell you if your head is tilted or not.

Why is it important for the brain to be able to tease out the difference between gravity and a linear acceleration? Dan Merfeld, director of Harvard University's Jenks Vestibular Physiology Laboratory, gives a graphic example: "If you imagine you're running through a forest—let's put ourselves back several hundreds of years—carrying your spear and dodging and ducking under trees, you need to keep very good track of where gravity is if you're going to stand up. You can't have your nervous system confused by the fact that at just this moment in time you're going to accelerate to the side to duck around this tree. You need to keep track of gravity and linear acceleration so that when you come out from around that tree you're standing up and not disoriented."

It's only been about twenty years since scientists discovered that signals from the eyes and the vestibular system converge in a place in the brain called the vestibular nuclei. More recently, proprioceptor signals from muscles were also found to connect there. The four vestibular nuclei, clusters of nerve cells that are part of the brain stem, are the sites of much of the actual processing of the three sensory inputs for balance. The other is the cerebellum, a part of the brain that coordinates and calibrates movement—and perhaps has a role in coordinating cognitive functions as well, a topic we'll explore in a later chapter.

Yet another vestibular ambiguity takes place in the semicircular canals, where rotational movement is sensed. These organs measure the motions that occur, say, when you nod your head to agree or shake your head to indicate "no." Three canals lie within each vestibular apparatus, which is about the size of a pencil eraser, one behind each ear. The canals are positioned at nearly right angles to one another. The easiest way to picture them is to imagine the corner of a box where three sections—two sides and the bottom (or top)—converge. The three semicircular canals are positioned as these three parts of the box are, and they measure rotational movement along those three alignments, like the axes of a three-dimensional graph.

Each doughnut-shaped canal, filled with a special fluid, has a small bulge at one end called the ampulla. This is where the rotational movement is actually sensed. Here a gelatinous membrane, called a cupula, extends all the way across the width of the ampulla, sealing it like a door. Embedded within the cupula are tiny hair cells that register even its most subtle movement. When your head is motionless, the fluid within the canals remains still, telling your brain that there's no movement. But if you were to, say, turn quickly to watch a wildebeest grazing on your left, the canals would move with your head, but the fluid inside them would tend to lag behind because of inertia. This delay would push fluid against the cupula, distorting it ever so slightly in the direction opposite to your head movement. The underlying hair cells would then sense this bending of the cupula and relay motion data to your brain.

The canals evolved to detect the sort of movement most humans experience while walking, jumping, running, crawling, or swinging from tree to tree. They weren't designed for passive motion induced by modern modes of transportation such as boats, planes, cars, and trains, for example. Because the canals can sense only acceleration, not velocity, they stop registering movement once a constant velocity has been reached. Why this happens has to do with the way the cupula works. Sensing rotational movement, the cupula moves in one direction and its hair cells fire off a signal to the brain. But once a steady velocity has been reached, the cupula reverts back to the neutral position. Now it's telling the brain there's no movement, but actually the motion hasn't stopped. Another layer of ambiguity occurs when the cupula detects a sudden decrease in rotational speed. Because deceleration is a form of (negative) acceleration, it is detected by the canals. During deceleration, the cupula, which had been in the neutral position, is pushed by canal fluids in the direction opposite the actual movement. It's been fooled again.

As if all these weaknesses and blind spots in the balance system weren't enough, proprioception is also an inadequate sensor on its own, according to Alain Berthoz in *The Brain's Sense of Movement*. Although it tells the brain how muscles are moving and where body parts are located in space, proprioception provides only information about relative positions—for instance, where your arm is in relation to your shoulder, or your leg in relation to your torso. It doesn't provide clues about the absolute position of your body in space, which is necessary for any kind of complicated motion. For that, the brain relies on the vestibular system, which serves as a sort of inertial-gravitational guidance mechanism.

When we're walking or standing on a hard, well-lit surface, information from all three sensory inputs is roughly equal. Vision tells the brain how the body is oriented in relation to things like trees or doorways, which are (usually) vertical.

(Have you ever experienced one of those tilted-room illusions in an amusement park? These attractions show how powerful an effect vision has on the brain, especially the visual reference of vertical. When everything in the environment is tilted uniformly—windows, doors, and floor—we have the perception that our bodies are tilted, too.) Proprioception provides information on the relative positions of legs and torso. The vestibular system sends signals about rotational or linear movement, or, just as importantly for orientation, whether there is no movement. But when the environment becomes more challenging—in rocky, uneven, or undulating terrain; or when the surface consists of sand, mud, or gravel; or if you attempt to balance on, say, a log floating on water; or if ambient light is poor or absent—the proportions of sensory inputs change. Under these conditions, vision and proprioceptive information becomes less useful in maintaining balance, and because gravity is the only steady, unchanging reference point the body has, vestibular information dominates. On stable, even ground, when the ankles flex it means the upper body is tipping forward or backward. But on unstable terrain, those same ankle flexes may be necessary just to keep the upper body vertical. The central nervous system compensates for this difference by relying more on vestibular inputs than ankle proprioception. In low light conditions, the CNS uses more vestibular and proprioceptive input than vision.

Resolving such ambiguities is just one of the reasons we have multiple sensory inputs for balance. The other is that it's a fail-safe system. If one goes awry, the brain can often compensate, sometimes extraordinarily well, with the other two. No other sense has this sort of built-in redundancy, perhaps because no other sense is as critical for survival. Those who are blind or deaf, or who cannot taste, touch, or smell, can usually manage, with a little help from their family or tribe. Losing any of the five traditional senses isn't necessarily life-threatening. But to lose one's orientation to the earth, one's sense of up and down, one's position on the planet, is to undergo one of the most profound disturbances a human can experience. It is doubtful that a human could exist in a persistently dizzy or otherwise unbalanced state for very long without losing the will or the wherewithal to survive. So evolution has given the brain an intricate adaptation system to safeguard this most fundamental of all the senses.

Chapter Four

How Balance Contributes to Survival

Remember the character Treebeard and his fellow Ents in J.R.R. Tolkien's *Lord of the Rings*? Well, if trees on earth had resembled tree beings in Middle-earth, with the ability to walk, talk, and wage war, our houses today probably would not be made of wood. For the trees would have seen the approaching loggers with their axes and saws in hand and would have fought back, tossing boulders and flinging branches, or retreated with thunderous steps. But trees, once planted, don't move around. Nor does any plant, aside from orienting toward light. "Plants don't have a brain because they're not going anywhere," says Robert Sylwester, professor emeritus at the University of Oregon and an expert on the application of brain research to education. "And if you're not going anywhere, you don't even need to know where you are."[1]

If it's true that brains became necessary when creatures, to survive, had to move, then the sense of balance, critical as it is for movement, plays a vital role in an animal's ability to pass on its genes to the next generation.

An illuminating example of this theory of why brains exist can be seen in the marine creature known as the sea squirt, or tunicate.[2] In its adult stage it looks like a small urn, filtering water through its mouth and eating mostly bacteria. Some species have a brief, free-swimming, larval stage during which they resemble miniature tadpoles, with head, tail, and fins. Inside the head is a primitive brain connected to a long nerve cord similar to the spinal cord of vertebrates. There's also a single light-sensing organ and a balance organ called a statocyst. Once the larva finds a good place to settle down, it anchors in, nose first, and goes through an astonishing metamorphosis. Within a span of about fifteen seconds, the creature consumes its own brain, eye, statocyst, and nerve cord. When a sea squirt transforms into its adult stage and ceases to move, those organs aren't necessary any longer. "Clearly, active movement is dangerous in the absence of an internal plan subject to sensory modulation," writes Rodolfo Llinás, a renowned neuroscientist from New York University, in his book *I of the Vortex*. "How far can you [walk] before opening your eyes becomes irresistible? The nervous system has evolved to provide a plan, one composed of goal-oriented, mostly short-lived predictions verified by moment-to-moment sensory input."

Until recently, most people believed the nervous system's role was simply to gather information about the outside world. But many scientists now think there's more to it than that. The model now in vogue is one of a nervous system that has evolved to sense the here and now for the purpose of simulating the future. By accurately predicting what's about to happen, an animal is better able to survive. "Prediction is the ultimate and most pervasive of all brain functions," says Llinás. To back up his argument, he points to the predictive powers of the remote senses for detecting predators. Vision, hearing, and odor detection, as predictors, are superior to touch and taste because they can function at a distance, buying more time for a creature to take action. "It is nice to be able to see that a threat is coming," he writes, "as opposed to having to feel it through one's outer being first, or perhaps to have to taste it, in order to register belatedly its arrival."

Llinás calls the senses "conduits from the external world to the internal world of the brain." They detect what he terms "universals," fundamental properties of Earth's environment that affect the survival of every creature. The first and most important of these universals, according to Llinás, were light, heat, and gravity. It was (and still is) imperative that every animal on the planet, from microorganisms to whales, have knowledge of these properties, and so the first senses were created to detect them. Further down the evolutionary pike, additional senses were developed to measure and monitor other properties like sound, odor, and texture.

Photo receptors—primitive eyes—told a creature where to go to find the shelter of shadows, for those who remained in the light were often eaten. Think of cockroaches scurrying instantly into the dark recesses of a kitchen when the lights are switched on. Temperature-sensing cells on the skin provided information about whether an environment was too warm or too cold. And the other ancient sensor was the one that detected Earth's gravitational force, enabling a creature to know up from down and where it was in space. This is the origin of the human vestibular system.

Statocysts, like the ones found in sea squirts, were the first gravity receptors. Fossil remains of sea squirts date back about 540 million years, which offers some idea of how old gravity sensors are. As hard as it may be to conceive, sea squirts are distant relatives of ours, as we both belong to the same phylum, Chordata, along with all other mammals, fish, birds, and reptiles. The common features of every creature in this group are a backbone-like nerve cord, gills behind the mouth cavity, and an anal tail. (You might wonder how humans can be members of this phylum because it's obvious we don't have gills or a tail. The answer is: these features are evident only in our embryological stage, disappearing before birth.) The sea squirt is the oldest and most primitive form of chordate, while humans are the newest and most "advanced."

Statocysts are little sacs equipped with a set of hairs and single or multiple bits of lime or sand, called statoliths. Depending on the orientation and movement of the creature, the statoliths stimulate the hairs, which in turn send signals to the brain. The engineering principle of the statocyst is remarkably similar to that of human otoliths, the component of the vestibular apparatus that detects straight-line acceleration and gravity. Statoliths and otoliths share the root *lith,* which is from the Greek word *lithikos,* meaning "of stone." The statocyst uses granules of sand or lime; otoliths (which you'll remember means, literally, "ear rocks") rely on calcium carbonate crystals. (Most creatures, humans included, manufacture their own stones. Others, like lobsters and crayfish, harvest them, replenishing their supply after they molt the lining of their statocysts. This change suggests that when animals ventured up off the ocean floor, from which they had an endless supply of otoliths, evolutionary forces conspired to create calcium granules to use as a substitute. Human otoliths may be modeled on sand, the legacy of a system refined millions of years ago on the bottom of the primordial sea.) The stones serve as little weights that bend the hairs when the creature moves in a straight line or as a result of gravitational pull.

The basic structure of statocysts remained unchanged until about 100 million years ago, when "modern" fish and reptiles first appeared.[3] As movement became more complex, the need arose for more precise sensors. That's probably why the semicircular canals evolved: having three canals on each side of the head provided more accurate information about rotational movement in each of the three dimensions of space: pitch, roll, and yaw. It is thought that the shape and function of the canals, which are basically hollow doughnuts, evolved from the structure of the lateral-line organ, the vibration-sensing tubes that run across both sides of fish and amphibians. According to biologists, there is very little difference, in form or function, between the vestibular systems of a shark, a timber rattlesnake, or a human being, a remarkable testament to the efficiency and utility of the design.

During the development of an embryo, the timing of a sensory organ's maturity not only suggests the evolutionary age of that organ—in other words, at what stage that organ appeared in its evolution over great expanses of time—but also may indicate its importance in subsequent neurological development of the organism. In the human embryo, the first sensory organ to be formed is the ear.[4] Just twenty-four days after the embryo is created, an indentation occurs in the ectoderm (the outermost layer of cells that will eventually become the skin and nervous system) called the otic placode. When the indentation becomes a little deeper it's referred to as the otic pit. Then that pit forms its own separate mass of cells, the otic vesicle, which now lies within the ectoderm along with the previously formed neural tube (which later transforms into the brain and spinal cord). Within five weeks, the two components of the inner ear, hearing and balance, begin to differentiate, with balance progressing at a faster rate than hearing. By eight weeks, both the cochlear (hearing) and vestibular (balance) components of the inner ear look much as they do in an adult, though smaller. Also highly mature at this stage is the nerve pathway that shuttles electrical signals from the vestibular apparatus to the brain, called the vestibular nerve. By about the thirteenth week, the vestibular nerve becomes "the first fiber tract in the entire brain to begin myelinating," according to Dr. Lise Eliot, a neurobiologist and professor of neuroscience at the Chicago Medical School. Myelin is a fatty substance that forms a sheath around nerve cells, insulating electrical signals that flow along the pathway and speeding up the transmission of those signals. It's sort of like lining the sleeve of a jacket with Teflon to help get your arm through it. By twenty weeks after conception, the vestibular apparatus "has reached its full size and shape, vestibular pathways to the eyes and spinal cord have begun to myelinate, and the entire vestibular system functions in a remarkably mature way," Eliot writes in her book *What's Going On in There? How the Brain and Mind Develop in the First Five Years of Life.*

But why is it important for a fetus, who must wait twenty more weeks before birth, to have a full-size, fully functioning inner ear? Is it for hearing or equilibrium? Studies have shown that a human fetus can hear low-frequency sounds like bass notes, along with the sound of its mother's voice and her gurgling digestive tract. But scientists don't know what purpose is served by this limited hearing ability. Perhaps it's sort of a preadaptation to language learning, but there doesn't appear to be any inherent survival value in it. More likely, it's the equilibrium sensors of the inner ear that are critically important, and Eliot cites two possible explanations. For one, the evolutionary age of the vestibular senses indicates that "the vestibular system is precociously poised to transmit sensation that is...critical to...early brain development." The other essential function of the vestibular system at this stage, says Eliot, is in allowing the fetus to know up from down. Why is that significant for a creature floating weightlessly in amniotic fluid? Because without this knowledge, it wouldn't know how to orient itself when performing a clever gymnastic move, a simple somersault, which allows it to get into the proper head-down position before birth. Studies have shown that babies with a malfunctioning vestibular system have a much greater chance of being born breech (buttocks or feet first), with all of the health complications that presentation can cause.[5] This ability suggests that the vestibular system may be an important contributor to a fetus "knowing" the best position to assume for a successful birth.

So if the vestibular system, the body's dedicated equilibrioception organ, plays an important part in human fetal development and orientation, what might have been its role in survival after birth, in children and adults? How might it have helped our ancestors live long enough to breed? To glimpse a possible answer to this question, let your mind wander back about 35,000 years to the continent of Europe, the scene of one of the great mysteries of anthropology: why Neanderthals, an early subspecies of the genus *Homo,* suddenly disappeared from the continent after living there successfully for nearly 200,000 years.

As scientists first began to ponder this enigma, their reasoning was influenced by a notable event that didn't look like a coincidence: the demise of Neanderthals occurred soon after the arrival in Europe of another species of *Homo, Homo sapiens,* who migrated there about 35,000 years ago from either the Middle East or Africa. The prevailing view was that Neanderthals, because they were more brutish, less intelligent, than Sapiens, were no match for the superior, big-brained newcomers. But a new picture has emerged, based on recent discoveries about Neanderthal's hunting style and body structure, including that of its vestibular apparatus.

In 2004, the British Broadcasting Corporation brought together a group of scientists who study Neanderthals, with the goal of trying to figure out the riddle of their disappearance. The first order of business was to build a complete Neanderthal skeleton, an important task given that only partial remains had ever been found, and no one had previously thought to create a composite skeleton by integrating pieces from various sites. Using bones from archaeological digs in France, Germany, and Israel, Gary Sawyer of the American Museum of Natural History spent a year piecing the skeleton together. His work then allowed other scientists to make some interesting observations. The skeleton of an adult male stood five feet four inches tall. Its sumo-sized rib cage and short limbs, with almost no waist, revealed that he was powerfully built. To sustain their large bodies, Neanderthals required a calorie-rich diet, which consisted to a large extent of meat and fat from animals. It is thought that Neanderthals lived on the outskirts of dense woods surrounded by tundra, hunting deer and other forest dwellers.

Evidence for *how* Neanderthals might have hunted also came from the BBC skeleton. Attachment points on the bones of the creature's right arm, and particularly the forearm and hand, indicated that muscles there were much more powerful than counterparts on the left side. The imbalance was so great that researchers concluded that Neanderthals must have thrust rather than thrown heavy stone-tipped spears into their prey (more advanced weaponry, the bow and arrow or spear-thrower technology, the latter consisting of a short piece of wood, called an atlatl, to which a spear was attached, greatly increasing its range and power, had not been invented yet). Using spears, they would have had to kill at close range, probably while lying in ambush as their prey walked along trails.

While this technique worked beautifully for something like 2,000 centuries, allowing Neanderthals to thrive in places modern human beings would find extremely challenging at best, it may have contributed to their undoing. About 45,000 years ago, the climate began to undulate through a series of rapid changes, from cold to colder, back to warm, then back to cold, sometimes shifting dramatically within a few generations of Neanderthals. The weather wreaked havoc on the woodlands where the Neanderthals lived and hunted. As the forests died off, so did the Neanderthals' primary source of food. At about this time, modern humans began migrating into northern Europe, bringing with them a different hunting technology: lightweight spears that could be thrown. It was a technique well suited to hunting on the increasingly open plains, for prey could be brought down at a distance, without the hunter needing to conceal himself. But these weapons required special attributes that Sapiens had but Neanderthals did not. There was certainly no lack of strength on their part; the skeleton's proportions proved that. And the two subspecies were probably on par in intelligence, judging from the cranial capacity of the BBC skull (Neanderthal brains were actually 20 percent bigger than a modern human's and equally sophisticated). It turns out that one of the biggest and most critical differences between the two was the size, and thus the sensitivity, of their vestibular systems.

By looking at the size of an animal's semicircular canals compared to its body size, biologists are able to make accurate assumptions about its agility and movement. Big canals mean more sensitivity to motion, and hence more nuanced control of bodily movement.[6] The larger the canals relative to body size, the more agile the animal. Birds, for instance, have relatively large canals, as you would expect in a creature that makes its living darting and diving and spinning through the air. On the other hand, a blue whale's semicircular canals are slightly smaller than those of a modern human; although the world's largest mammal is lithe and graceful, its movements are generally slow and methodical.

Fred Spoor, a biologist in the department of anatomy and developmental biology at University College London, has spent much of his career measuring the vestibular apparatus of a wide variety of animals, including an array of primates and fossil hominins. Making highly detailed X-ray scans (also called CT or computed tomography) of the bony labyrinth part of the skull, which houses the vestibular apparatus, he's able to determine the precise size and shape of its semicircular canals, even though they were composed of soft tissue that disappeared shortly after death. Charting these dimensions across hominin subspecies, Spoor found that as humans evolved, the size of the two canals involved in vertical orientation became larger and more sensitive.[7] The cause may have been greater demands on the balance system as hominins moved from a mostly tree-bound existence, to a mix of living in trees and on the ground, and finally to terra firma full-time. In other words, from quadruped to biped. It seems that balancing on two legs is more demanding than on four, which makes sense to anybody who has observed a toddler attempting his first unassisted steps.

When Spoor measured the bony labyrinth of the BBC skeleton's skull, he found that its semicircular canals were significantly smaller than those of a modern human, and even of earlier human ancestors, an apparent anomaly in the evolutionary chain. "The only logical conclusion," he told the BBC, "is that Neanderthals were less agile, and probably didn't include much jumping and running in their general behavior."

And here may be the key to the riddle. If canal size is related to agility, then there's probably a strong link between a human's large semicircular canals and one of the qualities that distinguish us from all other primates, as well as from Neanderthals: the ability to run long distances. While it's a bit of a chicken-and-egg exercise—did larger semicircular canals lead to more proficient running, or did the demands of running select for larger semicircular canals?—one probably cannot exist without the other. Modern humans, with their larger semicircular canals and dozens of other anatomical features that made them highly suited for long-distance running, were perhaps better able to adapt to the environmental changes occurring in Europe when they arrived. Neanderthals, limited by their heavy physique, short limbs, broad pelvis, and inferior balance capabilities, weren't able to hunt prey on the open plains the way humans could. They were relegated to the ever-shrinking forests, and gradually their populations shrank along with their habitat.

While obviously an aid to avoiding predators and, as the humans who lived alongside Neanderthals demonstrated, a key to survival, running may also have helped guide the formation and structure of our bodies, and spurred the threefold growth of our brains since hominins diverged from the apes. While these ideas are highly speculative, as are many hypotheses about human evolution, some scientists believe they provide an intriguing new view of life on the African plains and are worth investigating further. How could running—and an oversized vestibular system—possibly have done this?

Daniel Lieberman is a professor of anthropology at Harvard's Peabody Museum. In the early 1990s, he began collaborating with Dennis Bramble, a University of Utah biologist, to study how animals run. While testing pigs on a treadmill, the two scientists had an epiphany. "Pigs are terrible runners," Lieberman told the *Harvard Gazette,* "and Dennis introduced me to an esoteric bit of anatomy that explains why." All good runners in the animal kingdom possess something called a nuchal ligament, which attaches the skull to the spine and keeps the head from wobbling while running. Modern humans have it, as do gazelles, deer, dogs, rabbits, horses, and many other agile mammals. Lieberman and Bramble wondered if the presence of this ligament could be used to assess the running capabilities of our human ancestors. Since every skull, even those millions of years old, shows evidence of where the nuchal ligament attaches to it, scientists could use that as a marker for running ability. The pair began sifting through skulls at the Peabody Museum and came up with an interesting pattern. No signs of a nuchal ligament could be found in the skulls of primates or in those of the earliest hominins. Creatures who lived mostly in trees had little need of running. But the skulls of *Homo erectus,* our most direct and recent ancestor, who spent most or all of their time on the ground, did have it. Excited by this discovery, Lieberman and Bramble went on to look at other running-specific characteristics of humans and found several that could best be explained by their utility in running.[8] They had nothing or little to do with walking. These features included such things as a narrow pelvis, a long Achilles tendon and an arched foot for shock absorption and energy storage, and an enlarged and powerful gluteus maximus that fires predominantly during running. And to stabilize the head and body during running, according to Lieberman and Bramble, humans evolved enlarged posterior and anterior semicircular canals, as Spoor discovered.

Here's a scenario for how these evolutionary advances might have played out on the African savannah a million years ago. A baking sun overhead. Wide open grasslands punctuated by glades of trees. Three *Homo erectus* males in their early twenties peer quietly over the grass, scanning for prey. They spot a group of gazelles resting under a tree, sheltered from the noonday sun. The men whisper and gesture to themselves, crouching as they move through the grass, one man armed with a rock, the other two with six-foot spears. They emerge from the grass and walk toward the herd, nervous and watchful, constantly wary of lions. The gazelles see them instantly but hesitate several moments before springing into the open, floating through the air in giant leaps. The men don't even try to match the animals' blinding pace, but simply fall into a relaxed stride behind them, perhaps a shade slower than a modern marathon runner.

As they leap over rocks and fallen branches, sidestepping trees and brush, their heads remain almost motionless, with no discernible vertical movement, while their arms and legs churn rhythmically, effortlessly, beneath them. Alain Berthoz, the French neuroscientist, says that during running and other complicated physical movements, the head is stabilized in the same plane as the direction one is looking. "The brain uses the gravity detected by the vestibular system," he writes in *The Brain's Sense of Movement*, "to stabilize the head and create a mobile platform as a frame of reference." Similar to the effect of an airplane's gyrostabilizer, the vestibular system allows running humans to maintain their balance as they navigate around obstacles and over changing terrain. It also contributes to visual acuity while running, through the vestibular ocular reflex, which keeps an object in focus despite movements of the head. This reflex would have come in handy when trying to keep prey in view while bouncing across rough terrain or when throwing a projectile at a moving target.

Lieberman and Bramble believe that most scientists have never seriously pondered the role of running in human evolution because humans have always been considered such poor runners. When you compare sprinting speeds, that's certainly true. The world's fastest human is half as fast as a horse or greyhound or antelope at full stride. And the "metabolic cost" of running is greater for humans than for most other mammals of similar body mass; humans use twice as much energy to cover the same distance as a quadruped roughly the same size. But when it comes to endurance running—the pace of an elite marathoner or even a weekend jogger—humans aren't too shabby. In fact, over long distances, and particularly in hot weather, when our extensive sweat glands and lack of body hair offer superior cooling, we can outrun dogs, keep up with horses, and even catch up to fleeing gazelles.

Eventually. For this kind of hunting would take hours, if not days. The gazelles would likely sprint until they thought the coast was clear, then seek shade again, their bodies dangerously overheated. Following their tracks and maintaining a steady pace, the humans would close in and soon cause another mass exodus. This cycle might be repeated several times before one animal, perhaps weakened by disease or injury, began to lag behind the rest. Singling out this creature, the humans would follow it for a few more miles until it fell, exhausted and nearly dead already. Short work with the spears and rock would finish the animal.

Although there's no way of knowing exactly how our early ancestors hunted, a few clues exist in the historic record of indigenous peoples. The anthropologist Grover Krantz called this type of game pursuit "persistence hunting."[9] Although Krantz and others assert that it probably was displaced about 40,000 years ago when bows and arrows, nets, and spear throwers (atlatls) were invented, evidence exists of its recent use. The Tarahumara Indians of Mexico, who even today are famous for their long-distance running skills, used to chase down deer in the twentieth century, in one- to two-day hunts, "until the creature falls from exhaustion, often with its hoofs completely worn away," according to a 1935 account Krantz cited. Shoshone Indians of North America and the bushmen of Africa also engaged in persistence hunting in recent times, the latter "whenever scarcity of vegetative cover makes stealth and concealment difficult and the use of short-range poisoned arrows impractical."

Even if this form of hunting didn't always prove fruitful, Lieberman and Bramble believe that running would have helped early man reach the carcasses of dead animals ahead of other speedy scavengers, such as hyenas and canids. Seeing telltale signs of circling vultures, hominins would have known a carcass was nearby and run to find it before other animals had removed every morsel. Between hunting and scavenging, hominins would have reaped a rich reward of protein and fat "thought to account for the unique human combination of large bodies, small guts, big brains and small teeth," the scientists report in their 2004 study, which appeared in the journal *Nature*.

Krantz goes a step further by hypothesizing that the mental skills necessary to carry out persistence hunting—keeping the task in mind for several days and anticipating the result—would also place selective pressure on bigger brains. "As the reward in food for successful pursuit of game tended, on the average, to go to those individuals with the greater mental time spans," he wrote, "selective pressure would favor larger brains with better memories."[10]

While a larger and more sensitive vestibular system may have allowed early human hunters to run, jump, and change directions quickly while traversing rough terrain and to keep prey in clear focus while throwing a weapon, it may also have contributed to survival in another way. It appears to play a crucial role in a neurological system used by the brain to navigate back to a desired location, say, the home village after a long day's hunt. Charles Darwin puzzled over the phenomenon in an article published in 1887:

> With regard to the question of the means by which animals find their way home from a long distance, a striking account, in relation to man, will be found in the English translation of the Expedition to North Siberia, by Von Wrangell...He there describes the wonderful manner in which the natives kept a true course toward a particular spot, whilst passing for a long distance through hummocky ice, with incessant changes of direction, and with no guide in the heavens or on the frozen sea.... We must bear in mind that neither a compass, nor the north star, nor any other such sign, suffices to guide a man to a particular spot through an intricate country, or through hummocky ice, when many deviations from a straight course are inevitable, unless the deviations are allowed for, or a sort of "dead reckoning" is kept. All men are able to do this in a greater or lesser degree, and the natives of Siberia apparently to a wonderful extent, though probably in an unconscious manner. This is effected chiefly, no doubt, by eyesight, but partly, perhaps, by the sense of muscular movement, in the same manner as a man with his eyes blinded can proceed (and some men much better than others) for a short distance in a nearly straight line, or turn at right angles, or back again. The manner in which the sense of direction is sometimes suddenly disarranged in very old and feeble persons when they have suddenly found out that they have been proceeding in a wholly unexpected and wrong direction, leads to the suspicion that some part of the brain is specialized for the function of direction.[11]

It turns out that Darwin's hunches were correct, as they were in other arenas. First of all, vision is extremely important in maintaining a sense of where you are and where you're going, as you'd expect, but so too is what Darwin called "the sense of muscular movement"—what today we would term proprioception. And as for the special part of the brain "for the function of direction," he was almost clairvoyant, hitting upon what some scientists today call the brain's neural compass.

During the 1960s several scientists proposed that the vestibular system played a role in spatial navigation, the planning and execution of movement from one place to another. An Armenian physicist named Beritoff seemed to prove this hypothesis with a series of experiments with dogs. As Alain Berthoz relates in *The Brain's Sense of Movement*, Beritoff placed blindfolds on dogs and guided them along a designated route in his lab to a food cache. He then brought them back to the starting point, before dispersing them to other locations in the lab. Even blindfolded, the dogs easily located the food, no matter where they started from. But if not vision, which senses were the dogs using? Beritoff found that the animals could find the food quickly when either the auditory, olfactory, or tactile senses were disengaged. The only time the dogs had trouble locating the food was when both vestibular organs were deactivated. "Though he realized that the cerebral mechanisms called into play were extremely complex," Berthoz writes, "Beritoff was the first to demonstrate a possible role of the vestibular system in dead reckoning."

Berthoz himself went on to show how this mechanism might work. He devised an experiment in which a human subject, seated in a revolving chair, is asked to look at a target in front of him and memorize its location. Then the lights are turned off and in total darkness the chair is rotated. After stopping, the subject waits four or five minutes, and then, still in the dark, looks to where he believes the target is located. Most subjects could locate the target "with extraordinary accuracy," Berthoz says, indicating that the horizontal semicircular canals, the only sensory cue available to the subjects, are helping the brain estimate head displacement. The vestibular apparatus thus assists the brain in creating a cognitive map of the world and keeps track of the head's position on that map.

Jeffrey Taube, a neuroscientist at Dartmouth College, added more weight to the theory with his investigation of a certain type of neuron dedicated to directional orientation. Discovered in the early 1980s, these neurons are called head direction cells. Taube pointed me to a video on a colleague's Web site that provided a surreal and graphic illustration of what these cells do. In the bottom of a small tub, a black and white rat scurries around aimlessly, as if looking for a way out. Attached to his head is a device that looks like a weird sort of crown; it is a surgically implanted electrode that connects a single head direction cell in the rat's hippocampus to a computer, via a black wire that runs straight up from the rat and disappears offscreen. As the rat wanders around the tub, every time its head points to about nine o'clock, you hear a sound like radio static. That noise represents the cell's "firing." In every other orientation of the rat's head, the cell is silent, inactive.

"If you were at home and, say, facing northeast," Taube tells me, "the cells that are coding northeast are firing right now in your brain. And if you turned you'd have a different population of cells that are encoding in that new direction, and the other ones would turn off."

The firing of these cells isn't controlled by the earth's magnetic field, he says, so they don't work like a regular compass. But Taube and others have experimented extensively on what does control the cells, what causes them to "encode" in a certain direction. Vision, it turns out, is very important, because when a visual cue—half of the tub wall was painted black while the other half was white—was rotated, the firing of the head direction cells changed to stay in alignment with the cue's new position. Vision is part of what's called the "landmark" system of navigation. If you're on a boat and you want to maintain a course, you take a visual "fix" on a point of land or an island, and you keep the boat on course by referring to that landmark periodically. The reference point is external.

The other system of navigation that influences head direction cells is called path integration, in which you use internal reference points, ones that lie within your body, to measure your movement through space. The brain uses feedback from muscles (proprioception), vestibular input (rotation of the head makes the horizontal semicircular canals fire, which helps orient the brain's cognitive map, as Berthoz's experiment showed), and something called motor efferent copy, which is a fancy way of describing the mechanism by which one part of the brain sends a copy of a motor command—such as "move legs"—to other parts of the brain before it sends that signal to your legs.

Many animals, especially those who don't have abundant visual landmarks in their environment, rely heavily on path integration to navigate. The reigning king in this unofficial competition is the desert ant. According to Jack Loomis, a professor in the psychology department at the University of California at Santa Barbara, the ant, in its continual search for food, traverses wide swathes of sand, covering hundreds of meters of zigzag foraging in a single outing.[12] Yet when he finds a dead insect, the ant is somehow able to make a beeline (perhaps *antline* is a better word) straight back to its nest. Swiss researchers studying Tunisian desert ants in the 1980s and 1990s came up with a method of testing their navigation abilities. When an ant found a dead insect, the biologists scooped up the ant and transported it to another location. Back on the ground, which was marked with grid lines, the ant walked directly to where its nest would have been if the humans hadn't played their little trick. Although research has shown that the ant uses the sun's position to help determine its orientation, that alone isn't enough to send it on a direct path home. Nor does the smell of its nest have an "attractive" effect because the displaced ant takes the same path toward its nest's original location even if it's in a direction different from the nest's new spot.

So how does the ant accomplish this little feat? "In theory," Jeffrey Taube explains, "what the ant had to be doing is monitoring its movements through the environment continuously, from its initial starting point until it got to the food. Then it had to make the assessment of 'Where am I?' and 'How am I oriented?'" And based on that information it would know its spatial orientation. It would know two things: where it was located and what direction it was facing. Then if it knew what direction the nest was, it could compute a trajectory or a plan of how to get back to the nest." Many animals have this ability.

Among mammals, the golden hamster's homing behavior seems to show that it relies heavily on two of the balance modalities, vestibular and proprioceptive input. Experiments have demonstrated that after a hamster is lured away from its nest with food, it returns directly to it even when it can't see where it's going. This research indicates that "inertial cues and proprioception are sufficient for successful homing."[13] For humans, though, the role of the vestibular system is a little mysterious. Jack Loomis reports that it may be useful for short distances and time spans, but that errors begin to mount up rapidly, decreasing its directional accuracy. Vision, he believes, is of far greater importance.

Jeffrey Taube, like most neuroscientists, used to believe that vestibular inputs didn't have much influence on head direction cells. Visual cues appeared to be much more powerful. But that view changed after he and colleagues made an accidental discovery in the early 1990s. They wanted to investigate the contributions to head cell encoding of each of the three path integration inputs: vestibular, proprioceptive, and motor efferent copy. They already knew that while vision was the most important sensory cue, rats could still orient accurately in the dark for long periods using path integration feedback, although the accuracy was not quite so good and it "wore off" after a while. They also knew that vestibular inputs alone were not sufficient for rats to maintain their sense of direction. They had made this discovery by wheeling rats on miniature carts into a novel environment in the dark. This restriction of mobility forced the animals to rely on just their vestibular inputs, not vision or proprioception or motor efferent copy. Errors in head direction cell firing grew significantly; the animals were, in a sense, "lost." But when, prior to repeating the experiment, the rats' vestibular systems were decommissioned with chemicals, either permanently or temporarily, something unexpected happened. "Lo and behold," Taube says, "to our surprise we lost the whole directional signal altogether. We actually couldn't conduct the experiment because there was no directional signal to look at." The system had shut down completely, in spite of the availability of visual, proprioceptive, and motor efferent cues. This result still puzzles scientists but strongly suggests a vital role of the vestibular system in path integration.

So how does this finding apply to human navigation? Taube, a native of New York, describes a scenario in which a resident descends into the subway system of that city. When she goes underground, she knows her orientation, based on visual landmarks and perhaps the position of the sun. She knows the location of north and of the Empire State Building and lots of other places. But after spending time in the subway system—going one way for a while, then changing directions, perhaps switching subways along the way—and then reemerging onto the street, she'll probably be disoriented at first. In the subway, she was forced to rely on path integration inputs, and when she sees the Empire State Building again, it isn't where she thought it would be, based on those inputs. So she has to recalibrate her sense of direction according to what she sees. "You have that 'Aha!' experience: Oh, the Empire State Building is over there! And if I were able to record your [head direction] cells, you'd see all the cells shift back to their original direction." Taube says the landmark system (visual cues), along with the path integration cues of proprioception and motor efferent copy, control the direction in which head direction cells like to fire. "But for actually just generating the signal itself, our data would suggest that the vestibular system is required."

It's clear how this ability to navigate would have been extremely useful to our evolutionary ancestors. "I'm sure early on in mammals this system was very vital," Taube says. "Take a rat, or any animal that's preyed upon. You might be foraging out for food or water or a mate, and if you all of a sudden get a sense that there's a predator around, you've got to make a direct beeline back to your hole. You're not going to want to take a circuitous route. Certainly...any small navigating animal [is] going to need to know that to survive." Although Taube is speaking here of rodents and other small mammals, the same would apply to African hominins 2 million years ago, or Siberian natives in the nineteenth century. Fleeing from predators who were higher up on the food chain was no doubt a more common activity for humans in prehistoric times.

We've already seen how the neural compass would have come in handy in locating water holes or seasonal food sources, or simply finding one's way back to the home cave after a multiday search for food in distant territory—but it's also particularly important today to anyone whose ambition is to become a taxicab driver in London. Hundreds of would-be drivers must spend years memorizing circuitous, even labyrinthine, routes through the city before they can qualify for a cab license. Eleanor Maguire, a neuroscientist at University College London, has performed brain-imaging scans on London cab drivers and found that the volume of gray matter in their hippocampal region, which is associated with learning and memory, and is where many head direction cells are located, was greater than in age-matched controls.[14] What's more, the size of the increase corresponded positively with time spent behind the wheel, suggesting that the experience of navigating actually increases certain brain functions.

But not every modern navigational need can be met by our old-fashioned balance system. Although throughout human evolution the vestibular system has helped our species survive, under certain conditions, as we'll see in the next chapter, it can also cause our demise.

Chapter Five

"Ear Deaths" and "Graveyard Spirals"

Nobody knows exactly what happened inside the cockpit of John F. Kennedy Jr.'s Piper Saratoga in the minutes preceding his fatal crash off Martha's Vineyard in 1999. The National Transportation Safety Board's official accident report stated that there was nothing amiss with the plane or its navigation system. There was no fire, no airframe damage during the flight, no empty fuel tank, nor any signs of alcohol or drugs in the bodies of the deceased. So what was it that sent the plane and its three occupants plunging nearly straight into the ocean, on a clear, calm night in July?

Most experts concluded that there were two probable causes. The first was Kennedy's poor judgment as a pilot. The second, and the one directly responsible for the accident, was the deceptions caused by his vestibular system. Ninety years ago, aviation fatalities like Kennedy's were so common that they were given a name: "ear deaths."[1] What countless other pilots had learned before him, Kennedy learned the hard way:

the vestibular system by itself isn't adequate for maintaining balance in flight.

It didn't take long after the Wright Brothers made their first tentative forays into the air, in the early twentieth century, for pilots to learn that flying was an extremely hazardous activity. Crashes, injuries, and deaths in aviation's first forty years were sometimes caused by faulty or ill-designed equipment, but a large portion were the result of pilot error: poor judgment, bad physical coordination, and, sometimes, the brain being deceived by its own senses.

In the era before the widespread use of onboard navigational instruments, most pilots had no trouble keeping their planes going where they wanted them to go—as long as they could see the horizon or physical features, natural or man-made, of the earth. They flew "by the seat of their pants," which refers to the "feel" they developed for piloting a craft in the three dimensions of space. But as soon as darkness fell, or fog or clouds obscured their view of the earth, a pilot's senses could easily become confused. Aviators would fly into a bank of clouds, for instance, and come out the other end with their wings cockeyed or, in some cases, even upside down—all the while thinking they were level.

These failures happen because our balance system was designed for one thing: two-legged travel on Earth's terra firma. We are a terrestrial species. The adage that if man were meant to fly he would have been born with wings isn't quite complete. He would also have been born with a vestibular system programmed to work flawlessly in the air. On land, gravity acts on the body as a uniform, constant force. Our sense of balance relies on it as a reference point, as do the balance systems of virtually every creature on the planet. But when we venture into the air, gravity no longer seems constant; moving through the three dimensions of space, the brain can be led to misinterpret centrifugal force—the pull caused by turning and climbing and diving—as gravitational force.

You might have experienced this phenomenon while traveling on a commercial jet. You look up, say, from reading a book to glance out a window and notice that the plane is in a sharp turn. One wing is higher than the other, and perhaps you can even look beyond the end of the lower wing to see features of the earth. Yet you don't feel as though the plane were tilted on its side. Proprioceptors in your muscles and joints sense pressure on your back and bottom, as if gravity were still pushing against you normally. Since the plane is on its side, gravity should be pushing you not into your seat but toward the earth, which at that moment might be in the direction of the person sitting next to you, or the row of seats on the other side of the aisle. But the pressure you're feeling on your posterior isn't produced by gravity; it's made by the centrifugal force generated by the plane's turning motion. It's the same force that pushes lettuce to the perimeter of a salad spinner. You are sideways to the earth, but your brain doesn't perceive the change in position because centrifugal force feels like gravity.

Each of the body's three equilibrium components—vision, proprioception, and the vestibular system—is vulnerable to illusion during flight. But the most troublesome to a pilot are tricks that can be played on the vestibular system's semicircular canals. As you may recall, the canals measure acceleration, not velocity. That's a good thing in most respects because if they measured velocity our canals would constantly be firing. They would measure such things as the rotation of Earth beneath our feet (about 1,000 mph at the equator) and the movement of Earth around the sun (about 67,000 mph). But thankfully the canals register only changes in motion, and once the acceleration slows down or stops, the canals tell your brain that the motion has ceased. Which, during land-based activities, it usually has. But in the air, that may not be the case. During a long, smooth turn in an airplane, that motion may continue for a long time. After about twenty seconds, however, the little motion-sensing appendage within the canal, the cupula, fails to sense the slow acceleration and returns to the neutral position, falsely indicating to your brain that the turn has ended. According to the Civil Aerospace Medical Institute, in some planes it would be possible to perform a full 360-degree loop so gradually that, with your eyes closed, you would not be able to tell when you were upside down.[2]

Perhaps the worst vestibular glitch, as far as pilots are concerned, is that when a turn in one direction begins to slow down, it may have the effect of causing the canals to register the change in velocity, but in the opposite direction. This error occurs because the cupula, having returned to the neutral position after the initial acceleration of the turn, misreads the deceleration as movement in the opposite direction, sending false signals to the brain. In a spiraling dive, the pilot, if he's relying only on his vestibular system, will make wildly inappropriate "corrections" to his course. These "rotation illusions," as they're called, are okay as long as a pilot has vision to corroborate and fine-tune the signals from the vestibular system. The redundancy of the equilibrium system shows its value here. Errors from the vestibular apparatus are corrected by what the eyes see.

There's a special name for all the tricks the vestibular system can play on a pilot in the air: "spatial disorientation," literally, the inability to know up from down. It took several decades of research, and what amounted to a large-scale propaganda campaign aimed at pilots, before ear deaths began to diminish. But as Kennedy's death and those of hundreds of pilots a year demonstrate, the problem hasn't disappeared.

Because the physiology of the vestibular system was not well understood in the early twentieth century, the first aviation medical specialists—who today would be called flight surgeons—entertained some terrifically misguided notions about the organ's role in flying.

In 1912, prospective military pilots were given a bizarre test that purportedly measured their equilibrium. A candidate sat on a piano stool and was spun around for several minutes at high speed. "If he vomited, he was rejected," Dr. Isaac Jones wrote in 1937 in *Flying Vistas: The Human Being Through the Eyes of a Flight Surgeon*. "This was not so good—particularly as it was rejecting the normal! With his head wobbling around like that, it was natural and normal for him to be made sick."

As I explained in chapter 1, anyone with a normally functioning vestibular system can be made motion sick. The only ones who can't are those with disabled or diseased vestibular systems. To varying degrees among different people, when a mismatch occurs among any of the three components of balance, motion sickness ensues. In the case of someone being spun in a chair, the vestibular system, once the spinning has reached a constant speed, is signaling the brain that there's no motion, while the eyes detect movement. Doctors of that era mistakenly believed that if someone could be made sick on a piano stool it proved he had a dysfunctional vestibular system—the opposite of the truth. But besides not being a good test of the health of the vestibular system, the piano stool test wasn't even a good predictor of whether someone was prone to airsickness. Many candidates who passed the test went on to get sick at one time or another during a flight.

Several other balance tests were used by the military air corps at that time. A U.S. War Department document in 1912, cited by Harry Armstrong in *Principles and Practices of Aviation Medicine,* declared that

> the following tests for equilibrium to detect otherwise obscure diseased conditions of the internal ear should be made:
>
> 1. Have the candidate stand with knees, heels and toes touching.
> 2. Have the candidate walk forward, backward, and in a circle.
> 3. Have the candidate hop around the room.
>
> All these tests should be made with eyes open, and then closed; on both feet, and then on one foot; hopping forward and backward, the candidate trying to hop or walk in a straight line. Any deviation to the right or left from the straight line or from the arc of the circle should be noted. Any persistent deviation, either to the right or left, is evidence of a diseased condition of the internal ear.... These symptoms, therefore, should be regarded as cause for rejection.

While these tests may sound strange for a pilot to take, they do resemble a typical exam that a modern vestibular physical therapist like Karen Perz would use to test the dynamic balance of a patient. However, the tests don't identify the specific contribution to balance of the "internal ear," as they were designed to. A person could fail any of these tests and still have a perfectly functional vestibular system, if the other two components of balance weren't contributing as they should, or if, say, there was muscle weakness in the legs.

It was thought at the time that the vestibular system could be "trained" to improve its capabilities, and hence a pilot's skills. One method was to use a bizarre-looking device called the Ruggles Orientator, which looked like an oversized steamer trunk encircled by enormous gimbals. On top of the "trunk" was a cockpit where a person sat. The motor-driven gimbals allowed an operator or the student pilot to orient the contraption in any position a flier might find himself in, in all three axes. It was supposedly a key to "developing and training the functions of the semi-circular canals and incidentally...to accustom [pilots] to any possible position in which they may be moved by the action of an aeroplane while in flight."[3] What's more, by blindfolding a student while in the Orientator, "the sense of direction may be sensitized without the assistance of the visual senses. In this way the aviator when in fog or intense darkness may be instinctively conscious of his position." But this view of the importance of the vestibular system was, of course, more hypothesis than fact and was later reversed.

Vision's contribution to orientation, in pragmatic terms, was well understood by the pioneer pilots of the period. But science hadn't yet unlocked the mystery of why this should be so. Every pilot knew that if he wanted to live very long he could fly only during the day, and only in clear weather. Before about 1925, "nobody actually went into a cloud if he could possibly avoid it," Jones wrote in *Flying Vistas*. "In fact we always waited until the wind died down before going up—usually about 5 p.m. or about sun-up. Our pusher planes were barely able to maintain flying speed. They were unsafe except under the best weather conditions—and not safe even then! It never occurred to anyone to fly in bad weather. So at first we had no chance to study the problem of flying blind. At that time all we knew was that we must make an earnest study of the 'Ear and Aviation.'"

Flying during daylight hours in good weather wasn't much of a problem in the early days of flight, when airplanes were more of a novelty than they were useful transportation. But in the period during and just after World War I, the restriction of daytime-only flying proved vexing.

Transporting mail by air, for instance, was handicapped by this limitation. Mail pilots didn't use maps or compasses to travel cross-country. Flying only during the day, in good weather and at relatively low altitudes, they simply followed railroad tracks, noting their progress by reading the names of towns printed on water towers. The tracks were more than just a visual marker. As dusk approached, pilots were compelled to land, for safety. Then, on a prearranged schedule, they transferred the mail to a train, which would continue traveling through the night. In the early 1920s, the government decided to speed up cross-country mail service. They knew that if planes could carry the mail from coast to coast, they could gain several days on the train-plane system. So officials came up with a plan to illuminate flight paths for night travel, building beacons every few miles.[4] As long as they could see the lights, mail pilots could fly safely at night. The first route of this kind was built between Chicago and Cheyenne, Wyoming, a distance of roughly nine hundred miles. These two points were strategically selected. Their location allowed planes leaving in the morning from either coast to reach them by dusk. Once there, pilots could continue flying all night along the lighted airway and reach the other end by dawn, when they could safely resume their routes to the coast. Within a decade, some 18,000 miles of airway routes had been illuminated across the United States.

The one drawback to this system was foul weather. Heavy clouds, thick fog, driving rain, or snow—these conditions would ground a pilot day or night. Something more was needed to keep the mail planes in the air every day, regardless of weather.

And something also was needed to combat the appalling number of ear deaths among military aviators. Pilots needed a way to determine their plane's position in the air without using their eyes. Isaac Jones, who coined the term "ear deaths," and his colleagues set out to thoroughly examine how the brain senses its orientation in the air. The net result of their studies was that a healthy vestibular system was indeed important for a pilot's "feel of the ship," as Jones describes it, meaning his sense of the plane's motion as it floats along in the sea of air. Contrary to the opinion of many veteran pilots, however, no amount of training or experience or machismo would allow a pilot to continue flying when he lost visual contact with the earth.

That would take technology. According to Jones, the Wright Brothers were the first to invent and use instruments that informed a pilot of his plane's position in the air. In 1912, they used a simple piece of string eight inches long that dangled in front of a pilot's head. "So long as this string pointed directly at the pilot's nose, the ship was flying without slipping or skidding," wrote Jones. Two years later, the brothers invented two other orientation instruments, a "pendulum bank-indicator," which described how a plane was turning in the air, and a rate-of-climb indicator. But Orville and Wilbur "found it difficult to get even their own student pilots to use these instruments, because of the humiliation when other aviators would say, 'The students of the Wright Brothers find it necessary to use instruments in flying.' In other words, for many years instruments were not popular. Fliers took pride in scorning them."

The ultimate solution to the riddle of "blind" flying came from technology used by ships. The marine "gyrocompass," devised by a Dutchman in 1885 and patented in the United States by the American inventor Elmer Sperry, solved a critical navigation problem on the high seas: in a steel-hulled ship, especially one fitted with electric motors, a magnetic compass wasn't reliable, requiring cumbersome shielding to be effective. The gyrocompass, as the name implies, used a massive spinning wheel, electrically operated, that interacted with the rotational force of the spinning earth to maintain a constant orientation to the north-south axis. Sperry, along with his son Lawrence, who was the youngest licensed pilot in the United States at the time and an aviation fanatic, then dreamed up a way to miniaturize the gyro components so they could be used in an aircraft—but not to determine direction.

The younger Sperry built a machine whose basic concept could have been lifted from the pages of an otology textbook. For he imagined that if three independent gyroscopes could be oriented to one another like the three semicircular canals of the inner ear, then motion along those axes could be controlled automatically, without input from a pilot. In fact, he wasn't thinking of the structure of the vestibular system when he came up with the design, but of the three types of motion a pilot can control in an airplane, pitch, roll, and yaw. Those three axes are, however, roughly the same as those of the semicircular canals. The pitch of an airplane is analogous to the motion of nodding your head yes. Roll is movement along the horizontal axis, during which one wing tilts higher or lower than the other, the same motion as when you tilt one ear toward your shoulder. And yaw is movement along the vertical axis, as when you shake your head side to side to indicate no. Lawrence Sperry called his invention a gyroscopic autopilot.

When he finally worked out the considerable problems of how to harness it to the controls of an airplane, he traveled to Paris to unveil his creation—weighing just 40 pounds and measuring 18 by 18 by 12 inches—to the world.[5] The date was June 18, 1914, and the occasion was an air safety competition offering a $10,000 prize to the person whose invention was judged to have the best potential to improve flight safety. Appearing last among fifty-seven competitors, Sperry, just twenty-one years old and having earned his pilot's license only nine months earlier, took off from the airfield in a Curtiss C-2 biplane with his French copilot and his autopilot, or as the French called it, a *stabilisateur gyroscopique*. He made three passes along the Seine that day, but the third was the most spectacular, calling for exquisite balance on the part of both the men and the plane itself.

On cue, Sperry and his companion wriggled out of their cockpits and gingerly stepped out onto each wing, grasping the struts for support. The plane would lurch momentarily with the shifts in weight but always righted itself quickly due to the gyros in the autopilot. The two men waved nonchalantly to the stunned spectators and judges in the reviewing stands below. Later, after Sperry was awarded the top prize, he demonstrated to French officials how his invention could even allow a plane to land and take off without human hands touching the controls.

Though the gyroscopic autopilot wasn't widely adopted by airplane manufacturers until decades later, Sperry designed other instruments that allowed pilots to fly "blind" as early as the 1920s. One was called a turn indicator, similar in function to the one designed by the Wright Brothers, which displayed the direction (left or right) and rate of a plane's turning. It was considered *the* essential instrument for blind flight. Invented in 1917, the Sperry turn indicator was standard equipment on most large military aircraft by the mid-1920s. There was one monumental problem, however. Pilots had no faith in it or any of the several other orientation instruments introduced later, such as the artificial horizon, which tracks a plane's pitch-and-roll movements against a gyroscopically stabilized horizon. Like students of the Wright Brothers, they preferred to believe the illusion-inducing input from their bodies' own navigational sensors: the vestibular system. Many of them paid for this conviction with their lives.

Those with the most piloting experience, according to Jones, were the ones most difficult to convince. Once again, technology, combined with some astute psychology, came to the rescue. This time it was an upgraded version of the piano stool that had been used to screen out pilot candidates during World War I. Instead of attempting to make a man so dizzy he would vomit, military flight instructors began using the spinning chair in the late 1920s to demonstrate to their students how orientation instruments could outperform the vestibular system in blind flight.

This new version of the stool was the Bárány Chair, the very instrument that the Austrian otologist Robert Bárány had designed to test subjects in his research on the vestibular system. Looking like a slimmed-down barber chair, equipped with high-quality bearings that gave it an exceptionally smooth ride, it had been used in the Air Service for several years as a teaching device. A few minutes in the chair were enough to demonstrate how the semicircular canals could send false signals to the brain. With other members of the class looking on, a student would sit blindfolded and be spun around at various speeds. After a certain amount of time, the chair would come to a stop. The instructor then asked him if he was still moving. The student invariably would answer yes—but in the opposite direction.

One day, a lieutenant colonel by the name of William Ocker brought a portable Sperry turn-and-bank indicator with him to one of the Bárány Chair demonstrations. Isaac Jones describes its voodoo:

> The pilot is rotated—preferably in the presence of other pilots. He is turned to the right and then the chair is stopped. He calls out, "I am turning left, to the left." He is then turned very rapidly to the right and then slowly to the right. He may say, "I am not moving," or "I am turning to the left"—whereas all the observers assure him he is turning to the right. The pilot is then told to look into the instrument box. A flashlight in the box shows a turn-and-bank indicator and a compass. As he looks at the instruments he is again rotated. He watches the instruments while he is being rotated. He says, "I am turning right; the indicator also shows I am turning right." When the speed of the chair is slightly reduced, he will say, "The indicator shows I am turning right; my senses tell me that I have stopped," or "The indicator says that I am turning to the right; but I feel that I am turning to the left." It frequently happens that the pilot who has just had this demonstration will argue that his sensations are correct, in spite of what the instruments tell him. In that case it is helpful to have him stand by and watch someone else go through the same performance. As a rule a few such experiences in the turning-chair will convince the pilot that he can rely upon the

instruments. His thought then is "Oh yes, I have that feeling of turning, but I am not actually turning." From that moment *his problem is solved.*[6]

Even seasoned pilots, after going through such a demonstration, were amazed by the seemingly miraculous abilities of the gyroscopically controlled instrument. Suddenly it became clear how useful, and in many cases absolutely essential, orientation instruments could be to a pilot. The Bárány Chair exercise quickly became standard training protocol for military pilots (and is still used today by the air force). Next, pilots were given extensive training flying "under the hood." In a two-seater biplane, the student's cockpit would be sealed by a canvas cover so that he could not see out; behind him in the other cockpit was the instructor, who could take over the controls if the student faltered. The Ruggles Orientator and, later, the Link Trainer, a more advanced flight simulator, were also used for instrument training.

It still took many years for a majority of pilots to trust their planes' orientation instruments. Jones reported that some diehards would return them to the manufacturer as defective and were incredulous when technicians told them they were operating perfectly. "When you start to question the instruments, that's where it gets dangerous," says Colonel George Maillot, a retired air force pilot and pilot instructor who's had several scrapes with spatial disorientation. "You can fall into a trap, unless you're Steel Man himself, to the [input from the semicircular canals], and they start telling you things, and you start believing them, almost to the point where you feel like you're standing on your head. And then it's straight-down time."

A pilot's visceral distrust in his instruments may be a testament to the powerful, primitive need for the brain to believe what its senses tell it, and to disregard or at least downplay everything else. Perhaps that's hardwired as a survival strategy, forged in a world where we needed fast, nearly automatic responses to stimuli—such as discerning the barely perceptible breathing of an approaching predator and knowing it was time to flee. It's difficult for the higher parts of the brain to override what the lower brain perceives as accurate sensory signals of any kind.

The jet age has presented new problems for pilots. With their tremendous speed and acceleration, jets are able to create forces that can disorient pilots in ways never imagined by Sperry or Jones.

One such problem, called the G-excess illusion, occurs to the inner ear's gravity and linear force sensors, the otolith organs, which measure horizontal and vertical linear accelerations, including the force of gravity. Tiny crystals, the otoliths or otoconia, embedded in a gelatinous material, push against hair cells in response to linear movement. They also measure tilt. Imagine a business card coated with motor oil. A layer of beads is then sprinkled onto the card. The beads easily remain in place as long as you keep the card horizontal. But if you tilt it at an angle, say, forty-five degrees, gravity tends to pull the beads downward, toward the earth. That's sort of what happens in the utricle, the otolith organ that senses movement in the horizontal plane, when it's tilted: the otoliths are pulled downward. Even though they don't move far because of the sticky stuff that holds them, there's enough movement to give off signals to the brain that the head is tilted. But any time the G-forces on the utricle exceed 1 G (the force of gravity on the surface of the earth), strange things can happen. A modern jet pilot experiences these types of accelerations often, and the sudden surge of power can be so enormous that, even though the trajectory of the plane is perfectly horizontal, the otoliths are pulled backward severely on their gelatinous substrate. It's as if you took that oiled business card with the beads on it and pushed it forward quickly. The beads would all tend to move backward. The brain misreads these signals from the otoliths as head tilt. The pilot thus believes he is going up when in fact he is level. His natural response? To push the stick forward to lower his trajectory. When this happens in a low-altitude flight, it can result in something aviation experts rather morbidly call the "lawn dart effect."[7]

The most common vestibular illusion in jet flight is something called "the leans," which is also experienced by prop-plane pilots. It happens when flying in turbulent air, usually at night or "in the soup." A pilot may be concentrating on his instruments to remain level and straight, but momentarily looks out the cockpit. If the plane suddenly rolls or tilts when he's looking up, and then gradually returns to level, the semi-circular canals will register the sudden tilt — if it has a force great enough to trigger them — but *not* the gradual return, whose force is below the detection threshold. The brain then perpetrates an illogical hoax against the pilot, insisting for several minutes that he is still "leaning" over to one side.[8]

On his final flight after a twenty-year career in the air force, Lieutenant Colonel Gregory Davis was in the left seat of a two-seater T-37 jet when he came down with a bad case of the leans. He thought this was ironic because his job as chief of aerospace physiology at Sheppard Air Force Base in Texas had made him an expert on this sort of illusion. He taught not only pilots, but flight instructor trainees, and had experienced the leans many times before. "The flight was uneventful until we descended back into the weather [clouds] for the formation approach and landing," he wrote in a report for a U.S. Air Force Research Laboratory. Making several turns on the descent to the runway, watching the lead plane in his squadron to stay in formation, he wasn't able to glance at his attitude position instruments to stay oriented. "It was really amazing," he continues. "There I was, on the wing of an aircraft, and I had a really good set of the leans. I could see the sun peeking through the clouds above me, telling me my airplane should be right side up. No, that didn't cure the leans. I had [a copilot] sitting next to me to tell me what attitude we were in at any given time. No, that didn't cure the leans. I knew the somatogravic [seat-of-the-pants] sensation told me I was right side up. No, that didn't help either....It was so bad, I was sure we flew most of the approach inverted."[9]

Spatial disorientation continues to be a major headache for the military air services. While the overall aircraft accident rates have steadily declined over the last thirty years, the rate of mishaps caused by "spatial-d," as air force pilots call it, has remained unchanged. This intransigence is mostly due to the increased performance capabilities of modern planes, especially single-seat fighter planes like the F-16, which can generate tremendous acceleration forces that the vestibular system simply can't handle. As the performance capabilities of aircraft increase, so does the frequency of spatial disorientation. Between 1991 and 2000, according to Lieutenant Colonel Davis, spatial disorientation has cost the air force sixty lives and about $1.4 billion in aircraft. That's an average of seven fatalities and over $100 million every year.[10]

On the civilian side, accident and fatality rates have declined steadily over the past twenty years, including accidents due to spatial disorientation. The National Transportation Safety Board listed 1,614 general aviation accidents in 2004, with 556 fatalities. Of the 312 accidents involving fatalities, about 40 percent, or 125, were estimated to be caused, at least in part, by "continuation of flight into weather for which the pilot was not qualified," according to the NTSB. Which means a pilot without an instrument rating got caught in or thought he could deal with poor visibility. In other words, spatial disorientation probably reared its head.

And when orientation is lost, the odds are stacked against a pilot's survival. In 1983, the Federal Aviation Administration reported that in a recent five-year period more than five hundred spatial disorientation accidents had occurred, with a 90 percent fatality rate.[11] Going back a little further, a study done by the University of Illinois in 1954 was the first to investigate precisely how such accidents occur. Twenty pilots participated in the study. They ranged from nineteen to sixty years old, and their flight time varied from 31 to 1,625 hours. None had any instrument training. In the experiment, each pilot was of course accompanied by an experienced pilot-observer. To simulate instrument conditions, the subject put on blue-tinted goggles, which prevented him from seeing through the orange-tinted cockpit windows of the Beechcraft Bonanza. The observer then noted how long the student pilot was able to control the aircraft, and the consequences. Flying blind, with only their vestibular system and proprioception to guide them, pilots could maintain control for just under three minutes. At that point, nineteen out of the twenty pilots went into what is called a graveyard spiral. The plane begins to turn in one direction, which is inevitable under these conditions. It happens so gently at first that the pilot doesn't notice it because the rotational velocity is beneath the threshold detectable by the semicircular canals. As the turn continues, the nose of the plane angles downward, increasing the airspeed, which the pilot usually *does* notice. To slow down, he instinctively pulls back on the stick. But instead of decreasing the speed, as would normally happen, the turn tightens and airspeed spikes. Sensing the mounting speed, the pilot continues pulling back on the stick, worsening the situation.

A diving spiral, a rapid, dizzying corkscrew descent through the air, ends with the plane breaking up either in midair or on impact with the earth.[12]

Like the twenty pilots in the University of Illinois study, John F. Kennedy Jr. was not instrument-rated, though he had completed about half of an instrument course. (Only about 20 percent of general aviation pilots are instrument-rated in the United States.) Licensed just two years before, he was what is called a "lowtime" pilot, with about three hundred hours of flight time under his belt. This is considered a particularly dangerous period for pilots because they have the ability to get into the air but lack the wisdom and experience to handle bad situations. Although he had made the two-hundred-mile flight from Essex County Airport in New Jersey to Martha's Vineyard many times before, he had always taken a flight instructor with him when the weather looked questionable. But the weather that night was clear and unthreatening, with eight-mile visibility. So when an instructor offered to go with him, Kennedy declined.[13]

He and his wife and her sister took off in the Piper Saratoga at about sunset, staying low to avoid air traffic from New York's airports, then moving to a higher altitude, a little under six thousand feet, as they flew above Long Island. At this point, visibility was probably still good: the last vestiges of sunlight still shadowed the terrain, and Kennedy could see pinpoints of light along Long Island. But when he reached the eastern end of the island, he faced a thirty-mile stretch of the Atlantic Ocean, with no lights below to orient and guide him. Night flying carries its own set of hazards, and one of them is losing sight of the horizon in a dark sky. There was no moon that night, and haze obscured the earth, according to other pilots flying in the vicinity at about the same time.

After about one hour in the air, as Kennedy began his descent toward Martha's Vineyard, where the trio was planning to attend the wedding of his cousin the next day, Kennedy's brain began to succumb to spatial disorientation. He then had, as the 1954 study showed, about three minutes from the time he lost sight of the earth to the onset of the graveyard spiral. As he struggled to control the plane, pulling up on the stick to slow what his semicircular canals told him was an increase in his airspeed, the spiral only became tighter. The motion probably caused him to feel dizzy, which may have added to the panic and terror of the situation, preventing him from taking logical steps that could have saved his life. Experts say he could have done at least two things.[14] He could have gotten on the radio to air traffic controllers to ask for help, or made use of the sophisticated gyroscopically stabilized autopilot system the plane carried, whose basic design was identical to the device Lawrence Sperry invented in 1914. Those three gyros inside the autopilot would have then taken over from the three misfiring "gyros"—the semicircular canals—in each of Kennedy's inner ears. The device, experienced pilots explained, could have taken over the plane's controls and brought it to within one hundred feet of the runway at Martha's Vineyard.

Crash investigators later reported that the autopilot was in perfect working order. But it was never switched on.

Chapter Six

Tonic and Stimulant

At the time, it didn't sound like a preposterous idea. One day, alone in the basement of my family's suburban Seattle home, I imagined that it might be possible, after a bit of practice, to stand on top of a basketball. I was a reasonably coordinated, though still gangly fifteen-year-old who, in the previous few years, had tasted the thrill of balancing on snow skis and a metal-wheeled skateboard. Standing on a basketball didn't seem any more challenging.

After retrieving the family ball from the garage, an asphalt-scuffed Wilson that my brother and I had used to learn the game in our driveway, I placed my right foot on top, gingerly, and practiced stepping aboard a few times, my left foot hanging in the air alongside. When I finally managed to get both feet on the ball for a couple of seconds, I remember feeling a glimmer of pride and prowess at just about the same moment that the ball, without warning, spun out from under me. The next thing I knew I was flat on the ground.

Instinctively I had broken the fall with my left wrist, which would require bone-transplant surgery and four months in a cast. But that didn't dampen my spirits or my ambition for learning to ski. I had started the sport the year before and loved the sensation of flying across the snow, suddenly untethered from gravity, teetering on the edge of disaster. I couldn't bear the thought of missing the ski season, so I arranged with the orthopedic doctor to have the cast set around my wrist in such a way that I could still grip a ski pole. People who observed my crude skiing style must have wondered why I had such a stiff and ungainly pole plant on the left side, but nothing could deter me from heading to the mountains.

One of a child's rites of passage is to challenge and stimulate the balance system. Some rites are more hazardous than others, some more foolhardy. But there is an undeniable tingling thrill in balancing on a ball (or trying to), riding a bicycle for the first time, roller skating, jumping on a trampoline, skipping, spinning on a merry-go-round, or playing ring-around-the-rosy. What is it about these kinds of activities that attract children? Are they simply fun and playful and exhilarating, or is there a more serious "purpose" behind them?

To answer these questions, let's hop back into the womb for a moment. In chapter 4 I wrote about the early maturation of the vestibular system in human fetuses. In addition to aiding in that crucial flip a fetus makes, just before birth, to a head-down position, the vestibular system may also contribute, after birth, to something both Victorian insane asylum directors and new mothers crave: pacification. Just as spinning in a rotary chair calms even the most tempestuous lunatic, so do gentle rocking and bobbing motions soothe a distraught infant. Whether the baby's cuddled in a mother's arms or is lying in a cradle, the rocking motions form the basis of a technique as old, no doubt, as the human species. But how can we be sure that vestibular stimulation is responsible? What about the physical contact, for instance, that occurs when a mother bobs her baby in her arms? Or the physical warmth of a caregiver's body? Or the soft, reassuring words of comfort?

Lise Eliot, in *What's Going On in There? How the Brain and Mind Develop in the First Five Years of Life,* cites a 1972 study in which researchers, working with babies two to four days old, tested a variety of "soothing methods." The babies were (1) lifted from their beds and cradled in a researcher's arms (tactile, vestibular, and body heat); (2) given contact without being lifted (tactile and heat); (3) placed in an infant seat and moved up and down or side to side (vestibular only); or (4) stimulated with voice (auditory only). In the end, those methods that involved vestibular stimulation were more effective than those that relied only on touch or contact, the researchers found. "Thus," Eliot writes, "the newborns cried considerably less when they were picked up and held upright over the experimenter's shoulder (providing both contact and vestibular stimulation) than when the experimenter merely leaned over the bed and held them close but did not change their position. Even pure vestibular stimulation—rocking a baby in an infant seat, without any caregiver physical contact—was more effective than contact alone."[1]

And age doesn't seem to matter when it comes to the pacifying effects of mild vestibular stimulation. Ever wonder why rocking chairs seem so popular in homes for the elderly? Researchers at the University of Rochester did and decided to test their benefit to nursing home residents diagnosed with dementia.[2] In the late 1990s, the Rochester scientists chose twenty-five patients to rock in a chair for up to two and a half hours a day for six weeks, followed by a six-week period with no rocking, noting any changes in their behavior. "Right away," said the researcher Nancy Watson, "nursing aides noticed the most dramatic effect: the chair served to calm someone down when he or she was emotionally upset." And the longer the patients rocked, the greater the effect. Among those who rocked more than eighty minutes a day, such behaviors as "crying or expressions of anxiety, tension, and depression dropped," according to the report. The heavy rockers also requested less pain medication during the weeks they rocked, and less than those who didn't rock as long. There were signs that they improved their balance as well, another indication that rocking targeted the vestibular system. "It's been very well documented with infants that a gentle repetitive motion has a soothing effect. We've shown that the same is true in an older population that is emotionally distressed," Watson concluded.

It's likely that stimulating the vestibular system provides more than just pacification. It may also be important for "organizing other sensory and motor abilities, which in turn influence the development of higher emotional and cognitive abilities," Eliot argues. Evidence supporting this theory is found in a fascinating study published in the journal *Science* in 1977.[3] Reflex and motor skills were tested in a group of twenty-six "normal, preambulatory" infants whose mean age was seven months. After undergoing the standardized tests, a group of infants received sixteen sessions of semicircular canal stimulation, induced while sitting in the lap of a researcher who was spun around in a chair in a dark room. Each spin began with a rapid acceleration, followed by a one-minute period of constant velocity, and ended with an abrupt halt. The children were spun both clockwise and counterclockwise, and their head angle was changed so that each pair of semicircular canals was individually stimulated. "The magnitude of canal stimulation provided in this study is similar to that produced by the cessation of prolonged whirling enjoyed by older children on small manually propelled merry-go-rounds and in games such as ring-around-a-rosy," the report said.

Two control groups of infants either were held in the chair but not spun or were not handled in any way. Four days after the sessions were over, all the infants were retested by observers unaware of the group to which an infant belonged. The results were dramatic. Children who received vestibular stimulation had reflex test scores three times higher than the control group, and their motor skill scores were more than four times better.

One of the more persuasive elements of the study was that the researchers had the benefit of observing a pair of three-month-old fraternal (nonidentical) twins. They each had identical pretest scores. One was placed in the spinning group while the other was in a control group. After the study, "The control group twin was developing head control but had not progressed beyond motor behavior in the prone and supine positions. The co-twin in the treatment group had mastered head control and could sit independently."

So in the span of one month, vestibular stimulation had noticeably accelerated the development of the spinning group's motor and reflex skills. How did this happen? The researchers suggested that stimulation of the canals had matured the vestibuloocular reflex (the reflex motions of the eyes that compensate for head movements), which "in turn provided the visual system a more stable retinal image against which motor involvement with the environment developed more rapidly." They also believed that vestibulospinal reflexes, the ones responsible for maintaining posture, might also have matured more rapidly in the canal-stimulated group.

One delightful but unsurprising aside noted by the researchers was that the spinning infants seemed to relax noticeably during the experiment, some even falling asleep as they spun. After the first few days of spinning, many would "babble or laugh during the rotation" and in the thirty-second interval between spins would start to fuss and cry.

So we've seen that vestibular stimulation has at least two effects, pacification and excitation. Slow, controlled movements, such as rocking or bobbing, ease babies (and rocking chair riders) into a relaxed, calm state. As the pace or rhythm of the stimulation increases, there's an arousing effect that speeds up the development of bodily movement and reflexes. The area of the brain that may be responsible is called the reticular activating system (RAS); it receives nervous signals from the vestibular system and another part of the brain called the cerebellum, the coordinating center for both movement and, researchers have recently discovered, some forms of cognition. The RAS governs such things as the sleep cycle, turning off the "higher" areas of the brain, the neocortex. And it also "awakens" the neocortex, priming the nervous system for action, and possibly for learning and growth.[4]

This intimate connection between the vestibular system and the RAS is one reason vestibular stimulation is at the core of a controversial yet widely used therapy program for children with certain types of behavioral and learning disorders, called sensory integration. Anna Jean Ayres, a California occupational therapist and psychologist, created SI in the 1970s.[5] She believed that in order for a child's brain to function properly, it must be able to organize and integrate the bombardment of sensory input it receives during the first decade of life. Social and behavioral problems may arise when this process doesn't occur normally.

When the body and all the senses work together efficiently, adaptive responses to changes in one's environment (for example, walking on a tilted, irregular surface) and learning become easier. The first and most important sense to organize and integrate, according to Ayres, is the one that tells "him about his own body and its relationship to the gravitational field of the earth, and then these become the building blocks that help him to develop the sense of sight and sound, which tell him about things that are distant from his body."[6] She believed that the vestibular system was the "unifying system" for all sensory input, providing a "framework for the other aspects of our experience." Ayres's theory explains why children up to the age of four seem to have an innate craving for vestibular stimulation, to the point of providing it for themselves with such spontaneous behaviors as rocking, bouncing, swaying, and head shaking.[7]

Some children with problems such as autism and attention deficit disorder respond positively to treatment that involves stimulating the vestibular system. In SI therapy, children are placed in swings that allow them to spin at rates governed by the therapist. Or they're asked to move around the floor while riding scooter boards in a prone position. These movements activate the vestibular system to "quiet, stimulate, or organize a child's activity level," according to Ayres.[8]

Dr. Stephen Glass, a Seattle-area pediatric neurologist, has been treating children with developmental and neurobehavioral disorders for twenty-five years, including many with what he terms "sensory processing problems," such as autism, Asperger's syndrome, and attentional deficits. "Though sensory integration has not been scientifically proven," Glass said in an interview, "it is a descriptive term that is very, very meaningful. It resonates with parents when they hear it. I've had almost no parent who I've explained the sensory integration issue to who doesn't feel like I've just described their child." He explained what might be going on in the brain during vestibular stimulation. "You put somebody on a swing or scooter board or a rotation device of some kind and you spin the bejeebers out of the vestibular system. What have we just done? Well, we've provided more sensory input in a brain that may have altered sensory threshold, and thereby you've created more stimulus perhaps where it is lacking." Glass said this increased stimulation may enhance the processing pattern of nerve signals, altering the "quality, consistency, and control" of movement.

Glass has seen enough empirical evidence that sensory integration therapy works to give him confidence that its underlying principles are probably sound. One element that lends credibility to the therapy, he said, is that the effects of treatment usually persist after the therapy session is over. "You would think that [SI treatment] would just have a transitory effect and wouldn't last," he said. "But the lasting effect is remarkable." As an example, he described an offshoot of SI therapy called hippotherapy (*hippo* is Greek for "horse"), during which children with severe neurological disorders spend time riding specially trained horses, which provide a number of sensory inputs, including vestibular stimulation.

"Just from my experience and some data," he said, "kids can get on the back of a horse, they ride several sessions, four to six weeks. Not only does their postural control and their balance improve, but also their overall alertness, their communication abilities, and their seeming ability to process other information, not just motoric [movement] information. And so there's a spillover effect that goes well beyond just the vestibular or proprioceptive pathways. And parents will say this as well. Intrinsic sensory motor stimulation improves not just balance and motor control, but overall awareness, overall attention span, overall processing, overall language." Glass says that several National Institutes of Health grants are fueling research to study the effects of hippotherapy.

Just as Ménière's disease and other disorders of vestibular organs provide insights into how an intact human balance system works, sensory integration therapy's success in treating neurological and behavioral disorders sheds light on why vestibular stimulation during childhood may be critical to normal development. Some scientists believe its power to affect cognitive and behavioral abilities may have to do with all the connections the vestibular system has with the cerebellum. I'll explore this subject in greater detail in chapter 9, but, as I mentioned before, the cerebellum is now thought to be a center for coordinating both physical movement and the "movement of thoughts," according to Dr. John Ratey, associate clinical professor of psychiatry at Harvard Medical School. In *A User's Guide to the Brain,* he even suggests a possible link between people with autism, a disease that causes a lack of coordination in basic cognitive functions, and people who are socially awkward. Autopsies on the brains of autistic people show definite malformations of the cerebellum, he writes, and perhaps "clumsiness, shyness, [and] nerdiness may derive from less obvious cerebellar abnormalities. Indeed the 'social klutz' is just that, awkward, uncoordinated, out of step, lacking social graces, all of it driven by an inability to properly pay attention, share attention, and coordinate the many simultaneously incoming and outgoing signals."[9]

As children grow older, they find plenty of ways to continue stimulating their vestibular systems and challenging their balance. In parks and schoolyards, playground equipment is often geared toward repetitive swinging or spinning motions: swing sets, merry-go-rounds, or single-seat spinning devices, for example. Like sleeping bats, kids hang upside down from chin-up bars, tightrope their way across narrow beams of wood, or bob up and down while skipping rope. Bicycling, skateboarding, scootering, Rollerblading—all these activities hone and develop the balance system, helping it mature.

Some kids require even more of a challenge, more stimulation, and if they're lucky enough to attend Seattle's Sanislo Elementary School, the sky's the limit. Here, a nationally recognized physical education program makes kids so proficient in gymnastics, unicycling, juggling, and double Dutch jump roping that they probably really could run away and join the circus.

I visited the school one rain-besotted December day during a biweekly session of the Unicycling Club. The gym where the club meets isn't particularly big, but it makes up in its sheer volume of unusual athletic equipment what it lacks in size. Rock-climbing handholds cover the wall in one corner; long pegs around the perimeter store sixty unicycles when they're not in use; dozens of Swiss balls hang suspended in a wide-mesh net behind one of the basketball hoops; gymnastics pads stand against the far wall; boxes of tennis rackets and juggling balls rest against another wall.

Adjoining the gym is the office of the physical education teacher Sue Turner, the mind behind the school's unusual program. A serious, no-nonsense woman in her late fifties, with short-cropped brownish blond hair that's starting to gray, Sue began teaching at Sanislo thirty-four years ago. Soon after she began her career, she decided that something was fundamentally wrong with standard P.E. programs, with their emphasis on tedious calisthenics and competitive games. So, brainstorming with her husband, another P.E. teacher, she decided to switch the emphasis to unconventional activities in which students would compete only against themselves. Her goal was to get large numbers of kids involved, all on an equal footing athletically. What she needed, Turner says, was "something new and different and fresh, where everyone was a beginner and would be challenged."

At a quarter past three, just after the regular school day ends, the gym is a riot of movement and noise as at least fifty children warm up for their hour-long unicycling session. They ride in random patterns across the dark brown hardwood, perhaps the most abused gym floor in the entire state. Every few seconds unicycles crash into it, pedals banging with a *thump* on maple, as their riders scramble to stay on their feet. Kids are talking and laughing exuberantly and look as though they could continue this sort of free-spirited, formless play for hours. Every level of skill is evident here, from kids who can barely stay upright while clinging nervously to a wall or bars, to one older girl who, remaining in a stationary position on her unicycle, practices jumping rope, each twirl of the rope passing underneath her tire. Several kids somehow manage to mount "giraffe" unicycles that put them several feet above the ground.

Order is restored when Turner grabs the microphone and begins the official program. She commands her troupe to mount their bikes and begin riding in slow, controlled circles around the gym. Miss T., as the students call her, displays the personal warmth of a marine drill sergeant. She speaks in short, sharp, upbraiding tones, doesn't joke, and rarely smiles. But it's obvious the kids adore her. Without hesitating, they ask her to help them tighten the retaining bolt on their seats and extend their hands to her so she can support them as they ride.

After a few minutes of warm-up, Turner has the riders line up on opposite sides of the gym, six groups of three riders on each side. The drills begin. Riders from opposing groups ride to the middle of the gym, lock hands, pirouette around each other, then head back to where they started. First the right hand, then the left, then both hands locked together for the 180-degree spin. The degree of difficulty rises. Next the riders meet each other in the middle, join hands, and one continues forward while escorting the other one backward, all to the tune of the "Hokey Pokey," in which "you turn yourself around, that's what it's all about!"

Miss T. admonishes her riders as they work: "It's not about going fast, it's about being in control." "Heavy on the seat, light on the pedals." "Keep your seat post straight!" "Bravo, bravo," she says to one adept pair, clapping as she praises them. The last trick she has them do, which she calls "the granddaddy of them all," is sitting on the unicycle with their feet not on the pedals but directly on the front of the tire. Small, careful movements of the feet propel the unicycle forward. Everyone is required to have two spotters, one on each side. Nobody can do this trick by himself, though a couple of riders roll several feet with just one spotter.

When the session is over, many kids are reluctant to put their unicycles away. Several manage to sneak in a few more rides. But nobody gets the milk and Pop-Tarts that are being wheeled into the gym until all the cycles are put away. A handful of children have been riding their own unicycles, which remain at their sides like cherished family pets.

Turner says that the time it takes kids to learn to ride a unicycle varies. One child, within five minutes, was able to ride the length of the gym. Others can't do it after two years of trying. "It's how determined you are. How badly you want to do it," she says. These kids, all under the age of twelve, seem amazingly competent, but Turner explains that while the club meets infrequently, riders have the opportunity to practice every day, at recess and lunch. It's apparently a very popular pastime: of the school's 350 students, almost a third know how to ride unicycles, and many are also adept at juggling, rope skipping, and gymnastics, the other core P.E. activities at Sanislo.

But what if your children don't have access to such an innovative P.E. program? What sorts of activities should you encourage them to do to fire up the vestibular system? Although any movement that involves the head, no matter how subtle, activates the highly sensitive vestibular organs, those that have the greatest effect aren't particularly sophisticated, don't require unicycles, juggling balls, or any special equipment, and have been practiced by children the world over for thousands of years. Just letting kids play in the rough-and-tumble way that kids usually do may be all they need. According to Carla Hannaford, a university-level biology teacher and counselor for children with learning disabilities, parents should encourage their children to "explore every aspect of movement and balance in their environment, whether walking on a curb, climbing a tree, or jumping on the furniture." In *Smart Moves: Why Learning Is Not All in Your Head,* she writes, "Crawling, climbing, rolling, spinning, walking on uneven ground, skipping, jumping...and spontaneous joy-filled play specially stimulate and develop the cerebellar/vestibular system. Walking on low boards, or climbing a rope ladder, monkey bars, large rocks or trees requires an amazing amount of balance, thus activating the vestibular system."[10]

More "organized" activities for turning on the vestibular system include walking and hiking, especially on rough, uneven terrain (which challenges balance more than flat surfaces do). Bike riding, swimming, rock climbing, tai chi, and yoga are also beneficial, as are sports that involve complex movements, such as tennis, Ping-Pong, and soccer.

When playground swings or neighborhood walks lose their allure, amusement parks still beckon. Rides have evolved tremendously in the last half century, but they've always been rife with vestibular stimulation. On trips to Disneyland in the 1960s, I remember being spun and whirled about in an oversized Mad Hatter's teacup, a tame though still slightly nauseating ride, but the highlight of our family's visits was always traversing the Matterhorn, riding around inside the mock mountain on mock bobsleds on rails. There's just something about roller coasters that keeps people coming back for more, despite the fear and trepidation (or maybe because of it). Perhaps it's the thrill of flying through the air, tethered to Earth by the slenderest of threads, poised to break free of gravity at the tops of the hills. Or feeling the unnatural centrifugal forces wrestle with your body in the turns and then the weighted heaviness of gravity sinking you into your seat in the troughs. These rides, and most of the other thrill rides that you find in theme and amusement parks these days, have in common an abundance of rotational and linear motion that stimulates the semicircular canals and otoliths. Of course, people don't flock to roller coasters thinking, Gee, it'd sure be fun to have my semicircular canals stimulated today. But it's undeniable that most kids are drawn to movement that arouses the vestibular system.

Perhaps the ultimate vestibular thrill ride that a nonastronaut can experience takes place high above the Gulf of Mexico, where modified Boeing jetliners perform maneuvers that would make an airline president's hair stand on end. Tracing the same parabolic arc that a roller coaster takes, the planes sharply ascend at a forty-five-degree angle, then level off to begin steep, stomach-dropping descents. This motion causes so much vestibular stimulation and sensory conflict that motion sickness occurs in about half of first-time fliers. The result—something on the order of 285 gallons of vomit from the passengers on one such plane, over about a decade of flights—led to its infamous nickname, the Vomit Comet.

Since the early days of the NASA space program, the U.S. government has run these flights in order to train astronauts and conduct experiments. The official name for it is the NASA Reduced Gravity Program. When a plane, traditionally a converted Boeing 707, reaches the top of each parabolic arc at an altitude of about 25,000 feet, everything inside is suspended in a low-gravity state for about twenty-five seconds, before it plummets toward Earth. During these periods, passengers feel varying degrees of weightlessness, depending on the shape of the parabolic arc. If the pilot wants to duplicate the gravity of Mars, or the moon, or that of a space walk, he can do so simply by changing the plane's path through the arc.

Don Parker, the University of Washington balance researcher, conducted experiments for NASA aboard one of these 707s during the 1970s. He was attempting to prove the hypothesis that microgravity increases fluid pressure in the inner ear, which could lead to vestibular dysfunction. It was well known that astronauts in space often displayed symptoms such as puffy faces and postnasal drip because of a buildup of fluids that on Earth would tend to be drawn downward by gravity. So Parker devised a way to study guinea pigs in microgravity, measuring fluid pressure changes in their semicircular canals. Because of technical shortcomings during the experiments, his efforts didn't bear much intellectual fruit, but he did find the flights themselves "entertaining as hell." The planes, which he recalls as having a distinct putrid odor inside the cavernous cabin, would begin and end their flights at Ellington Air Force Base, near Houston. "You'd go up to twenty-five thousand feet," Parker remembered, "and fall like a rock for five or six thousand feet. I think we used to do fifteen parabolas to New Orleans and fifteen back, and if you could go out afterwards and eat at Petey's Barbeque, which was right across the road from Ellington, then you were tough."

Parker never got sick on the flights, nor did any of his guinea pigs. He would carefully prepare the animals, rising at three in the morning to anesthetize them as a precaution against the possibility. "I didn't know whether a guinea pig could be made sick," he says, "and I didn't know if they barfed. I had spun guinea pigs in a centrifuge up to, God, at the very highest we did them at four hundred Gs [four hundred times the gravity we feel at the earth's surface]. We'd put them in a body cast [before the centrifuge sessions]. Man, those guinea pigs, if you put them in a snug body cast, they'd do fine. Those damn guinea pigs within fifteen minutes [after the body cast was removed] were eating, and within a half hour they were trying to mate. They're pretty tough little creatures." Even so, the guinea pigs were apparently not invited to eat at Petey's, despite appearing to feel just fine after the flights.

The NASA planes were expensive to fly, so each flight was crammed with researchers conducting their experiments simultaneously. This made for some interesting plane mates, according to Parker. One flight was loaded with students from a Christian college in Florida, who were being used as human guinea pigs for an experiment testing anti–motion sickness drugs. The lead scientist in that group, a man named Ashton Graybiel, told Parker that he had chosen Christian kids because he needed subjects whose bodies were relatively free of chemical substances, and during the 1970s they were "the only ones he could find who could reliably stay off drugs." Part of the protocol of Graybiel's experiment was to have the students do "paced head motions" to evoke motion sickness. Parker explained that if you keep your head still in a zero-gravity aircraft, you won't get motion sick or, if you do, not severely. But if you start moving your head around, you have more problems. "These kids were so entertaining," Parker related with a puckish laugh. "They were up there moving their heads back and forth, and in order to keep up their spirits they'd be singing hymns like 'Nearer, My God, to Thee.'"

If astronauts, college students, and scientists, as well as the occasional guinea pig, could find such exhilaration flying in a

Vomit Comet, why not arrange for the general public to experience microgravity for themselves? That business idea percolated for a decade in the mind of Peter Diamandis, who is famous as the founder of the X Prize Foundation, which offered $10 million to the first person to fly a private spacecraft into space and back (Burt Rutan and Paul Allen won in 2004). So in 2004 Diamandis, an MIT-trained aeronautical engineer and Harvard-trained physician, launched the Zero Gravity Corporation, which offers flights to the public on its own version of a Vomit Comet. Based in Fort Lauderdale, Florida, the plane, a 727 cargo jet with a specially modified padded cabin, carves about fifteen parabolas per flight, at an altitude of between 24,000 and 34,000 feet. The first two simulate Martian gravity (one third that of Earth), followed by two lunar arcs (one sixth that of Earth), and the remaining ten are at maximum microgravity (space walk–level). Photographs of happy fliers on the Zero Gravity Web site show attractive men and women playing Superman as they launch themselves across the cabin in midair, grinning like six-year-olds on Christmas morning.[11] As for the less savory aspects of what could be seen as the world's largest roller coaster, company officials claim that the flight plans have been moderated to reduce motion sickness. "The bottom line is that NASA flies its airplanes using a very different flight profile," an official told MSNBC.com. "Ours is designed for the maximum amount of fun and enjoyment." Customers must harbor a deep desire for this kind of fun and enjoyment, for the price of a ninety-minute Zero Gravity flight, $3,000, works out to about $500 per minute of actual time spent in microgravity. But, hey, where else can a guy experience the kind of excitement and weird vestibular stimulation that astronauts can on a space shuttle flight?

The quick and not altogether ingenuous answer is under water. Astronauts, for many years, have trained for space walks by donning space suits and submerging themselves in specially designed water tanks. By using air or weights to neutralize their buoyancy, they can simulate microgravity quite accurately. So any scuba diver, for a fraction of the cost of a Vomit Comet flight, could get the same kind of flying, floating, weightless sensations as an astronaut.

In the end, perhaps one of the great appeals of pursuits like scuba diving, parachuting, jumping around inside a zero-gravity airplane, riding roller coasters, or even such larks as skipping or bouncing toward the sky on a trampoline is to feel the bizarre effects on our vestibular systems, attuned as they are to more mundane and predictable forms of motion. It may be part of our innate desire to explore and master gravity, to feel its influence, or the lack of it, in every possible way. A small group of humans relish taking this notion to the extreme. Driven by strong impulses, they engage in pursuits that you or I would find foolhardy even to consider. If you asked them if they enjoyed all the novel vestibular stimulation, they'd probably look at you blankly. But their unusual feats, as we'll see in the next chapter, demonstrate just how spectacularly balanced our species can be.

Chapter Seven

Extreme Equilibrium

You watch Unus standing on one finger and you think, "Look at such a fine, intelligent and excellent man making his living standing on one finger when most of us can't even stand on our feet."
— ERNEST HEMINGWAY, FROM AN ESSAY WRITTEN FOR THE PROGRAM OF THE 1953 RINGLING BROTHERS AND BARNUM & BAILEY CIRCUS

Measured by any test you might care to devise, profes-sional ballet dancers and ice skaters possess exceptional bal-ance. So do top-level skiers, surfers, and gymnasts, as well as seasoned players of every kind of ball game known to our ball-obsessed species. But there's no better place in the world to observe the full panoply of balance prowess—by humans as well as other creatures—than a good ol' fashioned circus.

At the Ringling Brothers and Barnum & Bailey Circus, clowns juggle while riding unicycles, dance on stilts, bal-ance baseball caps on their noses and chairs on their chins, or climb freestanding ladders that tip over while they're on top. The horse trainer–contortionist balances on a ring suspended far above the floor. Animals get into the act, too: a row of elephants all standing on their hind legs, with their front legs resting on the flanks of the animal in front of them; sheepdogs who walk backward on their two hind legs; horses kneeling down on one leg as if "bowing" to the audience. Most circus performers are balance artists almost by definition.

In fact, several founders of the earliest modern circuses were performers themselves whose acts highlighted their extraordinary balance. The circus as we know it today was created in England in the late 1700s by a twenty-five-year-old entrepreneur who had just retired from His Majesty's Royal Regiment of Light Dragoons. His name was Sergeant Major Philip Astley. As a dragoon, or cavalryman, he had won fame as a trick rider. Early in his career, while practicing with his regiment, he so astonished commoners who watched him ride that he was thought to be "the devil in disguise," according to one account. He was seen riding "full speed standing upon his horse and leap[ing] off and mount[ing] again without the horse slackening his pace." This test of balance was followed by one even more diabolical: as his horse cantered around in a circle, Astley stood "upon his head with his heels in the air."[1]

After retiring at a tender age, Astley began a second career as a showman, charging audiences a small sum to witness his equestrian prowess as he and his riding students performed on a small field in London. The popularity of his horse shows allowed him to move into ever more sophisticated arenas. As the money rolled in, these venues began to look as ornate and regal as opera houses, as though King George himself might be sitting in one of the several levels of balconies that surrounded the circular stage, on which Astley and his wife, also an accomplished rider, rode their steeds.

Engravings from the period feature horses circling the ring at a thunderous pace, chandeliers blazing overhead. A man directs the mounts with a stick or whip from the middle of the ring while riders balance, on one leg or two, on the horse's backs, sometimes with another person standing on the rider's shoulders. In addition to humans balancing on horses, horses also balanced on their two hind legs. Astley was allegedly the first person to teach horses to dance on two legs, which they did to music. One steed even learned how to "lift a kettle from the fire and arrange the tea things for company," while standing on two legs.[2] Astley was by all accounts a savvy businessman, and before his audiences could get bored watching trick riding demonstrations, he spiced his show with itinerant clowns, acrobats, jugglers, and tightrope walkers. Thus was born the modern "circus," which in Latin simply means "circle," a nod, it seems, to the circular equestrian stage of Astley's design.

For centuries, since at least ancient Greece and Rome, such performers had been mainstays of village fairs and harvest festivals throughout much of Europe. But until Astley's circuses, along with those of his many imitators, began proliferating throughout England, Europe, and the United States, those performers had rarely had steady gigs. In 1882, one American tightrope walker who did manage to find regular employment in a traveling circus persuaded his four brothers to join forces with him and start their own troupe. The Ringling Brothers Classic and Comic Concert Company began by entertaining Midwest audiences, then expanded into the Northeast, growing bigger and more sophisticated as it gained fame. By 1907, the Ringlings were powerful enough to purchase their biggest competitor, Barnum and Bailey, which they merged with their own show into one giant circus in 1919. By 1956, however, the show came under hard financial times and dissolved. It was resurrected that same year by a promoter who believed the show could go on if it used indoor arenas as venues instead of giant tents. The new formula worked.

Although no longer owned by the Ringling Brothers, the circus still carries the old familiar name: Ringling Brothers and Barnum & Bailey Greatest Show on Earth, and it was inside a large civic arena a day before the show's opening performance that I met Crazy Wilson. One of the circus's featured performers, Crazy Wilson, otherwise known as Wilson Dominguez, has traveled with Ringling Brothers for nine years. Born into a Venezuelan circus family, Dominguez, thirty-three, is a fourth-generation professional acrobat. Gracious and humble, he doesn't appear at all out of his mind. His build is that of a jockey, short, wiry, and not particularly muscular, and he wears his wavy black hair combed back. When asked if he had good balance as a child, he answers, in his broken English: "Yeah, I think everybody born with something. I think I was born—thank you, God, you give me something—with some idea for good balance, some idea for creating new acts. When I go high I don't have too many scare."

Dominguez, like most circus balance artists, feels a strong compulsion to create acts that no one else performs. It comes from a sense that audiences (and probably circus managers, too) might become bored if they see the same act twice, so the degree of difficulty or amazement has to keep notching upward. Dominguez has devised two acts that set him apart from other balance performers, and for which he earned his professional moniker.

One involves a contraption called the Wheel. It's a forty-foot-long, two-foot-wide metal structure, oriented vertically, that spins around on its axis. On one end is a heavy counterweight, and on the other is a circular cage, which rotates independently of the Wheel and is big enough for Dominguez to stand up in. He starts his routine inside the cage, which is open on both sides. As the Wheel spins around slowly, he maintains a vertical position by constantly walking, like a pet hamster exercising on a circular treadmill. After the Wheel is spinning at a moderate pace, he then crawls *outside* the cage and stands on top of it. With nothing to hang on to (and no safety harnesses or nets), he has to keep walking in order to stay upright. The most spectacular element occurs just before the cage reaches its apex forty feet above the ground, when Dominguez launches himself off the end, somersaulting into the air. He hangs suspended for a long second, weightless, then reaches out and grabs any part of the Wheel within his grasp. It looks spectacularly dangerous and it is; of all the circus acts he has ever performed, Dominguez has the most respect for the Wheel.

"First of all," he explains, "it is higher than the high wire [forty feet compared to thirty]. And when I do the somersault, I don't see nothing. I have to find something [to grasp], you know? When I do the somersault, I put my hand like this [he flails around in front of him]. 'Where is the wheel?' I am like a cat, I grab on anywhere. The people like it; that's the most important. I work very hard."

The high-wire act that Dominguez performs comes later in the show. He and another Latino start out by jumping rope on the wire, their feet barely lifting to clear the rope as it twirls around them. Then comes a little sizzle, a signature trick that allegedly no one else in the world does. A tiny trampoline is mounted to the middle of the wire. Dominguez jumps up off the wire, springs off the trampoline, and flies a few feet into the air before landing on the wire again. Then, taking up the long, weighted balance pole, he stands on the wire just in front of the trampoline. His partner leaps onto the trampoline, launches into the air, and alights on Dominguez's shoulders.

To me, having read about the sensational stunts that other high-wire artists have performed in the past, Dominguez's act seems almost obligatory. Feats of superhuman balance on a high wire? That's old-fashioned, passé. Bring on the MTV brand of entertainment, the quick-hit, thrill-a-minute, techno-heavy approach. But I'd be willing to bet that balance feats on the wire that made headlines fifty or a hundred years ago would have a similar impact today, if anybody would dare to perform them.

Even skipping rope on the wire, which is considered an advanced skill, has been more flamboyantly executed by past performers. Harold Davis, who went by the stage name the Great Alzana, was the Ringling Brothers star high-wire performer during the 1950s. Wire walking was almost a misnomer when applied to his performances. Although he might begin a performance with a simple warm-up walk on the wire, he would quickly graduate to more difficult maneuvers: after skipping the length of the wire, he might break into a series of big hops, then begin running. Next he would grab a jump rope and in midspan begin a routine that would be a challenge to perform on the ground. He would twirl the rope forward, then backward, then take big bounces off the wire as he did double and then triple loops. His signature trick, and the most dangerous one, was riding a bicycle across the wire while carrying three women.

Over his forty-year career as a funambulist (a word derived from the Latin *funis,* "rope," and *ambulare,* "walk"), Davis suffered two fractured backs and countless broken ankles, ribs, skull, arms, and legs. After years watching Davis perform, John Ringling observed that he was "the greatest and most foolhardy high-wire artist who ever lived." Ringling himself personally lobbied the New York legislature, in the early 1950s, to pass a law requiring safety nets for aerial acts working more than twenty-five feet off the ground. He did it for one reason: to keep Davis from killing himself.[3]

The fall that spurred Ringling into action likely was the one that occurred in 1949, toward the end of Davis's first season with the circus. It happened in Miami, during the bicycle act, which he performed then with his wife and two sisters.[4] As Davis sat on the bike with a balance pole in his arms, Hilda, his younger sister, stood behind him, her feet supported by metal pegs attached to the bike's frame. His wife, Minnie, and other sister, Elsie, hung vertically from cables attached to the two wheel axles, spinning slowly in midair. Always during this act, Davis's father, Charles, who had taught his children how to walk a wire when they were young, spotted them from fifty feet below. Unnoticed by the crowds that night, Charles moved along in the dark like a shadowy angel, as spotlights illuminated the performers above. Watching his family intently, Charles at first was unaware of the heavy rope that dangled from the top of the circus tent directly in the bicycle's path. Harold saw it the moment he launched the bicycle out across the wire, but it was too late. Unable to pedal backward, he couldn't avoid the rope. When he realized what was about to happen, Harold began frantically yelling to the ground crew, but the band was playing so loudly that nobody could hear him. Seconds after touching his eighteen-foot, forty-pound pole, the rope pushed Harold off balance. "We've had it, we're going!" he shouted to the others as he wrestled with the weight of the rope. Clutching the pole, Harold wasn't able to grab the wire to save himself as he normally would have done. He could only hope that Hilda, behind him, was able to snatch it as she fell. But somehow she missed it too, and together they began their awful descent. Harold said he could see his wife's face, "white as a ghost," as he streaked past her.

But Charles was ready. Crouching low, timing his move precisely, he thrust himself beneath his children's bodies to break their fall. Instead of losing their lives, Harold and Hilda broke their backs. Meanwhile, Minnie and Elsie remained hanging from the bicycle. Their weight anchored the bike to the wire, and they were soon rescued unharmed. Harold and Hilda took months to recover from their injuries. Charles suffered torn neck ligaments and was released from the hospital after four days. For the next ten years, he repeated his vigilant walk at every performance, prepared to spring again to save his children from death.

Davis had been hired by Ringling to replace Karl Wallenda, the most famous name in the high-wire profession. Wallenda, too, had disdained using a safety net. Both men believed the presence of a net made for sloppy, undisciplined performances. The fear of injury or death from a fall, they believed, compelled a performer to execute movements perfectly. They also thought they owed it to their audiences, who would be more intensely excited by a performance knowing a mistake would have morbid consequences.

Wallenda learned to walk a wire in Germany as a teenager. He came by this skill in a circuitous way. After learning rudimentary hand-balancing skills as a child, from his family, who were part of a long line of circus performers, Wallenda supported them during World War I by doing chair-balancing tricks on *Bierstube* floors. Next he had to figure out how to make a living in the shattered economy of postwar Europe. He kept practicing his chair handstand act, finally mastering handstands atop *four* stacked chairs, all except the bottom one balancing on two legs. But his audiences were so poor that he wasn't able to coax much change out of their pockets even with this balance tour de force. He took a job in a coal mine but found that he couldn't tolerate the wretched and dangerous working conditions and soon quit, vowing never to return. Somehow he had to find a job as an acrobat that paid a living wage.

He thought he'd hit the jackpot when he saw an ad in a national trade newspaper seeking someone to do handstands.[5]

But not just any handstand. It had to be performed while the man who placed the ad, Louis Weitzmann, a fearsome brute just released from a Russian prison, was lying on his back on a rope suspended sixty feet in the air. Weitzmann would point his legs upward, and his assistant acrobat would grasp his feet and push up into a handstand. The trick had never been done before, Weitzmann explained, but if they succeeded crowds would come from miles around and make them wealthy. Though Wallenda had confidence in his own

The three-person pyramid shown here was the Flying Wallendas' finale after the troupe decided that the seven-person pyramid was too dangerous. In this photo, taken in 1972, Karl Wallenda, who is sixty-seven, is about to lower himself into a chair balanced on a steel rod. Behind him is his twenty-one-year-old grandson, Tino, whom Karl had trained since he was a child. The other man is Luis Murrillo. Notice the extreme concentration on the face of each man, his eyes riveted to an object in front of him to give him added stability. After sitting on the chair, Karl would then rise up into a standing position. (From the archives of Tino Wallenda)

skills, he wasn't sure how he would perform so high up, or whether Weitzmann could be trusted. If either of them bobbled, it was Wallenda who would pay the price, as he would be launched into space while his boss could easily grab the wire to save himself. But Wallenda was in a bind. Because he had no money, he felt he had to take the job. Weitzmann at first gave him just enough to sustain himself: a cot to sleep on, a monotonous diet of bologna sausage, and a rigorous training routine. Wallenda later discovered a chilling fact. Weitzmann had already gone through fourteen apprentices. None had been able to perform "the trick."

Wallenda's fear of Weitzmann compelled him to train hard to prepare for the act. He already knew how to do any imaginable sort of handstand, but what he wasn't familiar with was walking on a wire, which was necessary for him to reach Weitzmann, who would be waiting for him in midspan, lying on his back. So his boss taught him the rudiments of funambulism. Starting on a low wire suspended a few feet above the ground, he learned how to balance with the aid of a weighted balance pole. (The pole, which usually weighs between twenty and fifty pounds, serves the same stabilizing function as the outstretched arms of a person struggling to maintain balance while, say, walking on a narrow curb. Balance poles are used almost exclusively on the high wire, never on a low wire, and acts that don't use them are, as a rule, more difficult than those that do.) Once proficient, Wallenda worked on wires set higher and higher, until his wire walking, and the act, were perfect. Audiences across Europe were transfixed by the sight of Wallenda's wobbling body as he perched precariously atop Weitzmann's feet high above the town, with no net to catch him if he faltered.

His apprenticeship served, Wallenda left Weitzmann to start his own act. He enlisted a brother and two others in what would become a seven-act circus performance. Several of the acts Wallenda created were variations of the chair-balancing trick he had mastered as a child, now transferred to the high wire. (The chair actually saved him from being devoured once. When he fell from a wire suspended over an open-roofed lion's cage, he instantly jumped to his feet and used the chair to ward off the big cat.) In one such act, with a female performer (Helen Kreis, who would later become his wife) standing on his shoulders, Wallenda sat on a chair delicately balanced on a steel bar yoked to the shoulders of two men who rode across the high wire on specially modified bicycles. When John Ringling saw this sensational four-person pyramid in Cuba in the 1920s, he hired the Wallenda team on the spot to join his famous circus in the United States, and the Flying Wallendas, as they came to be known, remained mainstays of the Ringling Brothers circus throughout the 1930s and 1940s.

Wallenda was recognized in his own time as an innovator among professional high-wire performers. The tricks he invented were such crowd-pleasers, however, that other high-wire artists often stole them, making it more difficult for him to secure work, as circus owners were always looking for completely original acts. So in 1947 Wallenda set about creating a routine so difficult that no one would dare try to copy it. That's when he came up with the seven-person chair pyramid, one of the most challenging balance acts ever executed. It consisted of three human tiers. The first was made up of two pairs of men fitted with special shoulder harnesses. Each man's harness held the end of a metal bar. Standing on these two bars were two more men, with a horizontal bar extending between them. And on top of that bar was—what else?—a chair balanced by the woman sitting on it. Each person held a twenty-foot-long, thirty-pound balancing pole. The group seemed to inch its way along the high wire as a single entity, which one observer compared to the movement of a giant slug.

The Wallenda troupe spent hundreds of hours perfecting this act, and it was an instant hit. Crowds loved it. The competition couldn't copy it.[6] The Wallendas did the seven-person chair pyramid act for many years without incident—and without a safety net, as Wallenda demanded. Sadly, almost inevitably, the pyramid finally did fall, in Detroit in 1962. The results were predictably grim. Three of the four men from the first tier fell to the ground. Two died that night. The third, Karl's son Mario, was paralyzed from the waist down for the rest of his life. While one man from the first tier somehow remained standing on the wire, Karl and his brother Herman, who had both been on the second tier, managed to grab the wire as they plummeted. The woman at the third, topmost level landed on Karl, and he held her as a makeshift net was assembled below, into which she was dropped. Karl suffered a cracked pelvis and a double hernia.

Although Wallenda promised his wife that he would never do the seven-person pyramid again, he went back on his word a year later when a television film crew wanted the troupe to re-create the act for a feature on the Wallenda family. He persuaded her that this would be the last time. Convincing the rest of the troupe was another matter. Wallenda argued that they would do the trick only for the camera. But that was just a ruse. He intended to resurrect the act and, once perfected again, take it on the road. Seventeen performances later, however, Wallenda retired the "seven," respecting the promise he had made to Helen.

Herman retired from the troupe at sixty-two. Karl thought about quitting the profession too. But he could not. "I get so damn lonely on the ground," he said when he was seventy years old. In 1972 he walked across a wire stretched over the field at Philadelphia Stadium, entertaining Phillies baseball fans. A few years later, he walked 750 feet above Tallulah Falls in northeastern Georgia, despite a vicious wind that threatened to topple him. He even did two headstands in midspan to demonstrate his mastery. By then Wallenda was something of an icon; Hollywood portrayed him in a movie starring Lloyd Bridges. But his granddaughter Delilah felt he "was becoming unstoppable in his compulsion to top even himself." He didn't seem to care how high the wire was positioned or what he walked above. "After fifty feet, it doesn't matter how high you are," he was quoted in an Idaho newspaper. "I'd walk a cable ten thousand feet high, if they could put it up. I fall fifty feet, I'm dead anyway." A short time later, in Puerto Rico, Wallenda fell to his death from a wire at the age of seventy-three. Though it was widely reported that a sudden gust of wind had knocked him over, his relatives claim that the wire had not been guyed properly. The fall had nothing to do with a lack of fitness or bad technique. It was, in a way, an appropriate death for a man who felt most alive when he was walking a wire and who at one time said, "Life is being on the wire; everything else is just waiting."

But not all acrobats expect to extend their careers into old age. "You cannot do this your whole life, because what I do now is very dangerous," says Crazy Wilson Dominguez. "Every act is very dangerous." He'd like to perform perhaps seven more years, until he's forty. Then, among other things, he'll have the freedom to play the sports he loves but cannot participate in, soccer, baseball, and basketball. An injury incurred while playing would prevent him from working, a risk he is unwilling to take now. The one sport he yearns for more than any other is surfing. "When I see that," he says with childish glee, "I go, 'Oh my God, that's easy for me because I have balance,' you know?" With his gift for balance, one might think Dominguez would have mastered the sport in a few days, when he was vacationing in Acapulco once. But it actually took him a couple of months just to stand up on a board. "It's different balance because the board moves," he explains.

Dominguez shrugs off the various falls he has taken. Only two resulted in injury, the worst when he was seventeen and still learning to ropedance. He broke both legs when he landed that time. Traditionally, learning takes place on wires erected a few feet above the ground, to minimize the danger. As the apprentice's skills and comfort increase, the height can be raised. The learning curve for wire walking, however, is long, steep, and irritating, according to Anne Weshinskey, who began training on a wire seven years ago, at the San Francisco School of Circus Arts. She now performs on a tightwire and teaches at the circus school when she isn't working at her regular job as a librarian. Learning is also rather monotonous, she says, requiring endless repetition before the body "gets it." I ask her if she teaches anyone with aspirations of joining a circus. "No," she answers. "Ideally you'd want to start with a child or teenager, and they just don't have the attention span or the concentration it takes to do it. You have to be really willing to be alone. Kids just don't want to practice because it's boring. Walking the wire is sort of like playing scales on the piano. Most parents don't force their kids: 'Now get out there and walk five hundred times across the wire.' It's also frustrating because when you first get on, you can't stay on the wire. All you're doing is falling off, falling off, falling off. It gives you this really bad feeling of frustration: 'Why can't I do this?' I mean, you see squirrels and things running along [telephone] wires, but they have a much lower center of gravity, four legs, and they're about half an inch from the wire."

Anne, thirty-nine, whose stage name is Winnie, performs a routine on the wire that involves jumps, kicks, skipping rope, doing splits, riding a unicycle, and climbing an unsupported ladder. For her, skipping rope is the most difficult trick. But just getting to the point where one can walk comfortably on the wire is the biggest initial hurdle, a feat that took her a year of steady work to accomplish. "It's good to have an idea of where your center of gravity is," she says of the learning process. "The way I learned this was to do squats on the wire and then come back up. And as soon as you squat down, everything comes into the center. You feel everything stop flailing and you're still. You can see where you're balancing from when you do that. Once you can walk, everything else is easy. It's meditative, because you have to focus on the other end of the wire. You can only look at one thing, so you can only think about one thing, and your mind gets in this zone of really focusing on what you have to do with your body."

I asked her about a tightwire feat I had heard about that seemed patently impossible: walking a wire while blindfolded. The nineteenth-century French funambulist Jean-François Gravelet, known as Blondin for the color of his hair, had supposedly done it, along with many other high-wire artists. From what I knew about the human balance system, I figured that something as difficult as walking on a wire would require visual input. In a memorable demonstration of this principle, a Russian researcher in the 1970s explored the role of vision in human balance in a novel way—by using circus acrobats as test subjects. There's no record of how willing these acrobats were to participate, but perhaps they had little choice, as Russia was a Communist country at the time. The acrobats were asked to stand on top of one another's shoulders. Under normal conditions, stacks of about six trained performers, depending on their skill level, can maintain their balance in this position for several minutes. But doing this isn't at all easy. Postural sway, the slight corrective movements the human body makes as it stands upright, increases with each acrobat in the formation, least at the bottom and greatest at the top. In the Russian experiment, the acrobats faced the added challenge of balancing in the dark. Under these circumstances, towers of more than two men could not stay aloft. With two "blind" performers, only two of the three sensory inputs for balance—vestibular and proprioception—were enough to maintain the tower. The addition of a third acrobat increased the sway to the point where all three inputs were required.

Wire walking blindfolded, Weshinskey tells me with a laugh, is a hoax. Some people can walk a short distance on a wire with their eyes closed, "but not the whole wire." If a performer uses a bag to cover his head, it's made of a porous material like burlap, so he can still see. If he uses a blindfold, he can still look down his nose at the wire. This subterfuge raised a red flag: were other seemingly difficult stunts on the wire rigged as well? Weshinskey confirms my suspicion, but quickly adds: "I'm not really at liberty to disclose those things." Then she backtracks a little. "Even the things that a performer tries to make look harder than they are are still really difficult. A back tuck [a backward somersault in the tuck position], the front somersault, there's no easy way to do them." She explains that the front tuck is one of the hardest things you can do on a wire because during the somersault you can't see the wire. You're facing up, toward the sky, and your feet land on the wire before you can see it. Few people in the world can do this stunt, and no more than a handful can do a double back tuck. One Chinese acrobat, working at Cirque du Soleil, can do a back tuck from one wire to another wire five feet above and three feet over from the first wire. This is incredibly difficult, Weshinskey says. Another person she's heard about can ride a giraffe unicycle on the wire, and perform a back tuck off the unicycle and land on the wire. Yikes!

Before she took up the tightwire, Weshinskey learned several other circus arts, including trapeze and hand balancing. The act requiring the most balance, she says without hesitating, is "tightwire, because it doesn't move. You only have the wire. You have to stay over the wire. You don't have any leeway to lose your balance. You have to keep it all the time, otherwise you're off."

Before a Ringling Brothers performance, as Dominguez and I talk, four female members of the Chinese acrobatic troupe practice next to us. It is utterly impossible not to be distracted by their grace and skill. Two of them quietly perform handstands on low wooden benches, handstands so exquisitely balanced and still, on two hands and one, that they look somehow inhuman. Dominguez notices my diverted gaze and looks toward them as well. "They are the best," he says. "Phenomenal. For me, the Chinese have the best balance in the world. With the hands, they are the best." Dominguez expects his preteen girls to follow in the family tradition and become circus performers. The daily presence of the Chinese acrobats provides an excellent role model for them, he says. The girls are learning contortion and hand balancing, as the Chinese acrobats have, and also basic trampoline skills, as Dominguez did when he was a boy. But the circus education the Dominguez girls are receiving probably doesn't begin to match the training the Chinese girls had in their native country.

Hand balancing is probably at least as old a circus art as funambulism. Standing or walking on one's hands looks bizarre and difficult, and that makes it a natural for entertaining crowds of people who could no more stand on their hands than they could eat with their feet. It's one of the core maneuvers of acrobatics, a performance art involving juggling, contortion, trapeze work, or wire walking. Its legacy in China goes back, some say, three thousand years. It is still considered a classical art form there, revered to the same degree that opera or ballet is in Europe. Touring professional troupes are not only a mainstay of Chinese culture but have become well known in the West as a component of various circuses. The origins of acrobatics in the Far East are murky. Some suggest it began when farmers met at fall harvest festivals and entertained one another with magic and simple acrobatic acts, such as walking on ten-foot stilts or spinning several plates in the air at once. Others believe it grew out of the practice of martial arts.

Hundreds, if not thousands, of children are schooled in the acrobatic arts today in China (and also in Russia, eastern Europe, and South America). Whether taught by expert parents or at special schools, such as the Beijing Acrobatic School, children's instruction begins at about age five. Much of their training consists of daily exposure to exercises that challenge the balance and muscular systems, such as walking on unstable objects like ropes or balls, performing handstands, contorting the body, and juggling. For the first couple of years, students learn the basics of balancing, tumbling, dancing, flexibility, and strength. Handsprings, somersaults, and headstands are mastered through hours of intense and often physically painful practice. Then student acrobats spend the next three to five years working on specific acts, almost all involving highly developed balance skills.

Some of the tricks don't look all that difficult until you think about trying to do them yourself. How about performing a one-armed handstand on your partner's head while both of you use your free arms to spin plates at the end of sticks? Or doing a handstand and hopping around the stage while every part of your body except your arms is squeezed inside a tiny barrel? Another eye-catcher is doing a one-handed hand-stand on a short pole, while your partner does a one-handed handstand with his hand resting on the back of your head. These are all part of a Chinese acrobat's palette of stunts.

The hallmark of any acrobatic routine is the performers' nonchalant, anybody-can-do-this persona. But this high level of proficiency takes years of intense practice. Most perform-ers don't take their acts public until they are teenagers, after they have truly mastered them. Two of the Chinese acro-bats in the Ringling Brothers show are cousins who perform together, Yongjie Yuan, a petite but incredibly strong girl of twelve, and Fei Yuan, a seventeen-year-old boy. Through their Chinese coach and a translator, I spoke with them, or attempted to, as we sat near the perimeter of a circus ring be-fore the evening performance. Both children were dressed in sweat clothes. I was struck by how childish and frail Yongjie seemed. She had a distant, bewildered look in her eyes that made me think she should still be at home with her family, not working three hundred shows a year in a strange coun-try. Neither one spoke or understood English, and neither did the coach, who insisted on answering all my questions, through the translator, even though I tried to direct them to the kids. During the interview both children seemed remote and bored, more interested in the two Jersey cows who were circling the adjacent ring, commanded by a young dark-haired woman. Fei sometimes held his face in his hands as if he were exhausted. About the only time they looked in my direction was when the translator relayed my questions, al-ways to the coach, an irritable man who seemed annoyed by my presence.

He confirmed that Fei and Jie, as she is called, had started their training when they were six or seven years old, but had trained together for only the past two years. They were part of a troupe of fourteen from China, eleven acrobats plus the coach, an academic teacher, and the translator. The children had learned the basics of acrobatics in a special school in Henan Province, in central China, where they trained five to six hours per day, four days a week, with two days devoted to academic work, and one day of rest. Children are often encouraged by their parents to audition for the school, for if accepted they can bring prestige and honor to themselves and their families. Acrobats are as revered in Chinese society as movie actors and pop stars are in the United States. But the social cost to the families—and to the children—can be steep. Once accepted at the state-run schools, often located far from their hometown, children rarely get to see their families again. Perhaps the heavy sadness I saw in Fei and Jie was part of the toll for being one of the planet's most skilled balance artists.

At one point in the interview, the coach said something to Jie, and she popped into a handstand on a small platform, curling her legs and spine into a severe C shape. Then she proceeded to do body presses in this position, bending her arms as she lowered her body toward the floor. This feat takes enormous strength, yet she didn't look remotely like a bodybuilder. She knocked off several presses in a row, and I wondered aloud how she could do this. The coach shrugged. I curled my arm into a strongman's biceps flex, pointing to her. For the first time, she giggled.

At the end of the interview, I asked the kids if they would demonstrate some tandem balance moves. Prompted by the coach, they took off their sweatshirts and Fei tossed his arms around a few times to warm up. Then he grasped Jie's hands and hoisted her easily over his head, where she did a handstand on his upraised arms. Then he proceeded to lower her down and press her up again several times, another impressive display of strength. Then Jie did something that stunned me. She transferred one of her hands to the top of Fei's head, then released her other hand from his grasp: a one-handed handstand supported only by his head. Fei made subtle corrections in balance with his neck and shoulders to keep her weight directly above him. His neck strained somewhat at the weight, but she held perfectly still. Then, after fifteen seconds or so, he reached up for her hands and down she dropped with a thud.

Later, when I watched them in an actual performance, Fei was more active than he had been in front of me. With his sweats off and in an armless tunic, he looked much more muscular than he had before. As Jie maintained a handstand on his upstretched arms, he went from a standing position to a crouch to lying on his back on the floor, and then up again. I was so riveted by their performance that I barely managed to look at the acrobats performing in the other rings. I was struck by how few people in the audience applauded after the Chinese performed a balance stunt, as if no one could grasp the sublime difficulty of their moves. Or perhaps it was just the opposite: they were so dumbfounded by what they saw that a reverential silence was the only possible response.

In the hour before the performance began, I had watched another gifted hand balancer in a preshow ritual called Circus Adventure, where kids get to meet and mingle with the performers. He was a clown named Gabor Hrisafis, a dwarf born in Hungary who stood a hair under four feet tall. Dressed in his full clown outfit, he deftly grasped the low wire, about three feet off the ground, and jumped up into a handstand, which he maintained for about fifteen seconds before dropping back down.

Hrisafis was no ordinary balancer. For one thing, he had recently set a world record for the longest vertical jump in the handstand position. While working for the Moulin Rouge in Paris, as television cameras whirred, he pushed up into a handstand on a special platform two meters high, with twelve steps leading down to the ground. He then launched off the platform, flying over the steps, and landed on the ground on his hands. It was hard to believe that his shortened limbs could absorb the tremendous shock of a leap like that. But when I first met him, he was wearing a T-shirt and shorts, revealing the ripped musculature of a prize fighter, thick through the shoulders and arms, with beefy thighs. I asked him what sort of weight-lifting routine he did to have developed his body to such an extent, and he replied that he does nothing beyond practicing and performing his circus routines, which involve a lot of hand balancing.

Harold Davis, another small man with a sterling physique, was known for performing handstands on the wire—with *one* hand. He had mastered handstands as a youth, becoming so adept that he could hop down flights of stairs using only one arm. A photograph of Davis demonstrates his phenomenal power and balance: Wearing only shoes and long pants, Davis is doing a one-armed handstand on a rope strung low between two circus wagons. He is perfectly still, his free arm outstretched, his shoulder muscles taut and as hard as an oak burl.

Like Wallenda, Davis survived his entire decadelong stint with Ringling Brothers. Many other wire walkers had similarly long careers and lives. Blondin, the first person to wire-walk across Niagara Falls, in 1859, worked until he was seventy and died in retirement five years later. During the same period that Blondin was making international headlines, a Canadian named William Hunt, using the stage name the Great Farini, tried to steal some of Blondin's glory by parodying the master's feats over Niagara. After Blondin lugged a stove to the middle of the wire, 190 feet above the river, and cooked an omelet on it, Farini hefted a hundred-pound washing machine across, stopping long enough to clean a few handkerchiefs. Hunt, after a long career as both a wire walker and a trapeze artist, died of pneumonia at the age of ninety. Before him there was Madame Saqui, another great French funambulist, with whom it was said Napoleon was in love. Famous throughout Europe in the mid-nineteenth century, she fell on hard times late in her life and was compelled by poverty to ropedance for audiences at the age of seventy-one.

What is it about wire walkers that allows them to make careers out of an activity that would land most of us in the hospital within seconds, and do it several times a day, day in and day out, for decades? And then maintain their catlike agility and number of "lives" well into their old age? Do they offer any insights into how we ordinary people can achieve and sustain better balance?

The biographies of circus acrobats, especially Wallenda and Davis, provide a few clues. First, most of the performers started young, anywhere from five to ten. The story goes that when Davis was practicing in his backyard, walking on a low wire suspended over the family's garden patch, the earth was littered with squashed vegetables. Of course, early and intense practice is probably the best way to learn any physical skill. Kids ought to get out and find different ways to challenge their balance. What they learn as children can form the basics of skills they can carry into their adult lives. But, importantly, the body has an amazing capacity to learn new things at almost any age. Older folks may take longer to learn new balance skills, but the payoff will come with improved equilibrium in any activity and greater security from falls.

Second, most circus acrobats were highly motivated to learn. In China, success in acrobatics holds the promise of financial security in a country beset by poverty. For both Wallenda and Davis, the alternative to being a professional acrobat was working in coal mines, which both men considered far more hazardous duty. Once they became accomplished wire walkers, neither Davis nor Wallenda, at least early in their careers, could afford to fall because neither used a safety net. Their concentration and focus were heightened by the overwhelming desire to stay glued to the wire. Granted, few people, other than those who plan to pursue a career in acrobatics, have this kind of motivation. The rest of us can stay focused on our less ambitious balance activities by considering the gracefulness and improved athletic performance we'll gain from them—or the consequences in old age of *not* sticking to them.

Both Davis and Wallenda used devices that gave them more stability as they performed. Wallenda relied on a weighted balance bar, which pushed him down into the wire more firmly and also increased what physicists call his rotational inertia, the tendency of an object to resist rotation, in this case the rotation of a performer's body off the wire. Rotational inertia increases in proportion to how far the total mass of an object extends from its center of rotation. The farther the weighted ends of the pole are from the center of rotation (the performer), the harder it is for the performer to fall. The same thing happens when, losing your balance on uneven terrain, you instinctively reach your arms to each side, perpendicular to the rest of your body. You're improving your stability by increasing your rotational inertia. Hiking poles and canes can act as balancing devices in much the same way, with the added advantage of giving you another leg to stand on, another point of contact with the earth.

Another stabilizing technique many wire walkers employ is simply taking small steps rather than big ones. This gait maintains their center of gravity over their feet, the most secure position. Any time your body mass isn't over your feet, the odds are higher that you'll tumble. Watch an infant who has just learned to walk, or an older person whose equilibrium is challenged, and you'll often see them taking small, shuffling steps. They both instinctively know that this puts them into a position of good balance. Top-level tennis players understand too. The most agile and fleet-footed players take small, controlled steps as they move, so they can be in a balanced position that allows them to hit the ball with maximum control and power. When your equilibrium is compromised, by old age, rough terrain, a physical disorder, or even a heavy backpack, take small steps to stay balanced.

Perhaps the most salient lesson that circus acrobats have to teach is the necessity of practice. Not just the hours put in to learn the skills in the first place, but the discipline throughout one's life to keep body and mind tuned to the rigors of balanced movement. Wallenda practiced his art nearly every day, and he kept his supreme balance skills until his last wire walk in 1978.

Harold Davis had even greater longevity. After retiring from all public appearances in the mid-1970s, he still practiced regularly on the high wire in the backyard of his Sarasota, Florida, home. Each time Davis stepped out on the wire he risked his life because even as a senior citizen he refused to use a safety net. Davis continued walking the wire every day until doctors diagnosed him with a heart condition at the age of seventy-nine and then ordered him to stay on the ground.

Even so, according to his wife, in the final three years of his life he would climb up one of the support towers and just sit there watching the wire, awash in reverie. "I guess in his mind and his heart he didn't want to quit," she said.[7] The Great Alzana died of natural causes in 2001, at the age of eighty-two. In the next chapter I'll discuss ways to uncoil your own inner Alzana.

Chapter Eight

The Wallenda Within

Fidel Castro gave a textbook demonstration of a fall on October 21, 2004, when he tumbled to the floor after giving a speech at a public event in Santa Clara, Cuba. He was seventy-eight years old at the time. Television cameras caught every nuance of gravity's wrenching grip on the man. Freeze-frames were published in newspapers throughout the world.[1] In the first photo, he is confidently striding from the podium toward the audience, head erect, looking out at a sea of applauding admirers. In the next frame, it's obvious that he hasn't seen, directly beneath him, a small step down. In fact the step is barely visible in the photograph. Instead of anticipating it, Castro extends his leg as if the surface were level. The third photo shows him spilling onto the floor, breaking his fall with his left knee and arm. Afterward he announced on state television, "I'm all in one piece."

Gravity, the mutual attraction among all things, is one of the four known fundamental forces of nature, along with electromagnetic energy and the strong and weak forces that hold atoms together. Yet most of us tend not to notice gravity much, at least not in our day-to-day lives. In this way it's similar to the sense of balance itself: always present but rarely entering our consciousness. Gravity is an invisible, mostly benign force—unless, for instance, you're climbing a ladder to paint second-story window frames, or decide to take the stairs instead of an elevator. Then you begin to sense the power of this unseen energy, or at least to ponder its potential. For if by chance you happen to disengage yourself, even momentarily, from the intricate dance your body must perform with gravity—the yin-yang, push-pull gravity tango—you put your life in its hands.

Maintaining an upright posture, while appearing effortless, is actually an enormously difficult and challenging task. It requires the central nervous system to constantly monitor hundreds of muscles and nerves and make continual adjustments to maintain our center of gravity over our feet. Add movement, even something as "simple" as walking, and the challenge increases. The action of walking has been described as controlled falling. You pick up a leg, lean your torso forward, and "arrest" the fall with a step. Then you do it again and again, each step the start of a descent toward the earth. Locomotion brings more muscles and nerves into play than simple standing, and vision also assumes a bigger role. Without your being aware of it, your eyes track environmental markers that are perpendicular to the earth, such as poles, buildings, and trees. These cues inform the brain about how to orient the body. (In fact, let's return to an example I gave in chapter 3: the brain is so hardwired to take body orientation cues from the vertical position of objects around us that it's prone to the fun-house illusion of the "tilted house." If all the doors, windows, and corners of a room are tilted a certain number of degrees off true vertical, the brain gets suckered into believing the eyes and actually tilts the body to match what it sees.) Input from the vestibular system is integrated as well, communicating information about head movement and linear acceleration, in addition to keeping objects in visual focus as the head moves.

Monitoring all these inputs requires the equivalent of a very large and sophisticated supercomputer because the consequences of losing one's balance can be so dire. At 9.8 meters (roughly 30 feet) per second, the speed at which gravity "pulls" objects toward the center of the earth, even uncontrolled drops from heights as low as a few feet can be dangerous.

Like Castro, I experienced my own awkward tangle with gravity a few years ago, during an off-trail hike with my uncle in mountains near Sedona, Arizona. While descending a steep gully, I stepped on a loose rock. Suddenly my feet shot out from under me, as fast as they had when I tried to stand on a basketball thirty-five years before. This time, instead of landing in a heap on concrete, I began sliding downhill at an alarming velocity, my arms trying to grab onto any passing object that might slow or stop me. After skidding down a near vertical cliff for about thirty feet, my left foot hit a ledge and I launched into space. Reflexively putting my arms in front of my face, I landed on my chest on a slab of rock. Gravity wasn't through with me yet. My momentum continued to carry me downhill, though at a slow enough rate that I could easily grab a tree to end this little joyless ride. Looking up to see the route of my fall, I considered myself incredibly lucky. I came out of it with only a sprained ankle and a broken rib, able to limp and shamble back to the car and eventually to a medical clinic.

I was fifty years old at the time and hadn't taken a tumble like that in thirty years of hiking. Was it just bad luck or did it have something to do with my now middle-aged body? Most authorities believe our ability to maintain equilibrium peaks in our twenties and then slowly begins to deteriorate until we reach our sixties, when it plummets. These statistics make sense when you consider that for most of anatomically modern humans' 100,000-year existence, our average life span was somewhere between twenty and thirty-five years. So balance skills reached their apex during the prime of life, giving our ancestors the greatest chance for survival until their genes were safely passed to the next generation. How rapidly the sense of balance diminishes depends on our genes, on the natural process of aging, and also to a large extent on how physically active we are and the types of activity we do. Barring a disease that affects balance, like Parkinson's, multiple sclerosis, Ménière's, or bilateral vestibular disorder, most reasonably active people can maintain their equilibrium without much effort until about the age of sixty-five.

All three primary sensory inputs for balance begin to atrophy in middle age. People who enjoyed good vision during their younger years usually require reading glasses, while others need bifocals, and still others suffer from macular degeneration or cataracts. Proprioceptors on the bottoms of the feet lose sensitivity, causing a lag in communication with the brain about the foot's position on the earth. And the tiny hairs within the vestibular system's semicircular canals and otoliths also lose sensitivity, decreasing the speed at which the gravity and motion sensors relay information to the brain. There is also a link between muscle strength, especially of the legs, hips, and trunk, and balance. If muscle mass and tone decrease significantly, as they do in most people as they age, then the body cannot respond as efficiently to sudden changes in equilibrium, such as those caused by slippery or uneven terrain. Without regular muscle-building exercise, strength levels decrease by about 12 to 14 percent per decade, starting at about age sixty in men and about age fifty in women, says Ben Hurley, a professor in the department of kinesiology at the University of Maryland.[2] And because older adults are the least physically active age group, this problem is particularly thorny.

With all these declines, it's a wonder we can stay upright at all past our sixties. In fact, many people don't. According to the Centers for Disease Control and Prevention (CDC), one in three Americans sixty-five and older falls each year,[3] which today translates to about 10 million falls annually. Falling is an alarming and growing health problem among the elderly in this country, a modern scourge that's only now beginning to get the notoriety it deserves, from both the general public and family practice doctors. It's a numbers game, and the odds aren't good. Consider that in 2003, according to the CDC, 13,700 Americans over sixty-five died from fall-related injuries, mostly from head injuries, and nearly half a million seniors were hospitalized due to a falling injury. While motor vehicle accidents constitute the leading cause of accidental death in every age category under seventy-five, falling takes over the number one spot after seventy-five, as people drive less and fall more. Some of the more famous seniors in recent years to have died from a fall were Katharine Graham, the eighty-four-year-old owner of the *Washington Post;* Dr. Robert Atkins, seventy-two, the creator of the Atkins Diet; and the television news anchor David Brinkley, who was eighty-two. Although those 13,700 deaths make up a very small percentage of the 10 million falls by the elderly each year, injuries from falling can also have a devastating result. Of those who fall, 20 to 30 percent sustain "moderate to severe injuries such as hip fractures or head traumas that reduce mobility and independence, and increase the risk of premature death," states the CDC.[4] That's roughly 2 to 4 million people a year.

Why older people fall is an enormously complicated question. The National Safety Council lists twenty-five different risk factors among four categories: medical, behavioral, environmental, and psychosocial.[5] Often a fall results from not one but a combination of factors. In addition to muscle deterioration, the list includes stroke and chronic illness, poor illumination and slippery floor surfaces at home, and anxiety and depression. If a person takes four or more medications, or any of the psychoactive meds such as antidepressants, the risk of falling rises. Poor vision and impaired cognitive abilities also contribute. Although it is rarely mentioned in the literature, obesity is probably also a major contributor, as added fat around the torso makes an already unstable body form even more unstable. Another factor that receives little notice is the rise in type 2, or adult-onset, diabetes, with its frequent symptom of foot neuropathy, a decrease in proprioception and tactile sensitivity.

Perhaps the most disturbing of all the risk factors is a psychosocial phenomenon known as fear of falling. This happens to someone who has already been injured in a fall and becomes so anxious about repeating the agony that he or she rarely ventures outside the home (although this reaction doesn't make logical sense, as most falls, at least among the elderly, occur inside the home). These people become more sedentary, which in turn leads to a greater chance for another fall to occur.

The published photos of Castro's fall don't offer evidence about any physical or mental deficits that may have contributed to it, only that his gaze was directed out to the crowd, not to the ground in front of him. It is possible that his close-up vision wasn't sharp, making it difficult to discern the subtle visual cues of the step's presence. (Bifocals, if he had been wearing them, would definitely have been a risk factor.) One thing an observer notices is the near instantaneous reflexes of his arms as the fall begins. As soon as his upper body lunges forward, the vestibular system sounds an alarm that kicks out his right arm into an almost horizontal position, where it tries to stabilize him like the balance pole of a wire walker. Of course in this case, the effort is too little too late, for his forward momentum cannot be halted. The left arm, meanwhile, moves reflexively into a position to absorb the initial impact with the floor, protecting Castro's head and chest. Milliseconds later, the right arm and left knee also move forward to help absorb the force. The strategy works. His skull remains intact. But his left knee and arm are sacrificed in the process, breaking under the load.

Among older adults, 87 percent of all fractures are due to falls, including the most serious type, hip fractures.[6] The bone that usually breaks is your thigh bone (femur), right at the top, where it angles into the hip socket. This area is crucial for bipeds. When you break this bone, the pain is so intense that you can't stand. Hospitalization is almost always required, usually for about a week. Doctors attempt to stabilize the injury by using surgical screws to hold it in place. In some cases they have to replace the femur head with a high-strength metal device that inserts into the hip socket. Over 300,000 hip fractures occur each year due to falls, says the American Academy of Orthopaedic Surgeons, at an average treatment cost of $27,000 per incident.[7] Even worse is the lifestyle cost. About half of older people who fracture their hips "cannot return home or live independently after their injury," according to the CDC. That means being forced to live with a relative or friend or at a nursing home. Only one in four people recovers fully. Half will require a cane or walker. And one in four will die within twelve months due to complications.[8]

Falls and falling are closely linked to nursing home stays. Even a person who falls once and isn't hurt is three times more likely to go into a nursing home than someone who has never fallen. And if a fall does lead to an injury, the person is ten times more likely to enter a long-term care facility than a nonfaller.[9] About 40 percent of all nursing home admissions are due to injuries resulting from a fall.

And for adults younger than sixty-five? For them, falling deaths usually happen on a job site or during recreation; a construction worker slips off a beam or tumbles off a ladder, or a rock climber loses his grip on a cliff face. In 2002, 3,413 people in this age category died as the result of falls,[10] most of which can usually be chalked up to bad luck or a lapse in judgment, not to a physical problem.

Falls occur with surprising frequency across all age groups. They're the second-leading cause of unintentional injury death in the United States (16,200 in 2003), after motor vehicle accidents but ahead of poisoning, choking, drowning, fires, and suffocation. Although the elderly make up the vast majority of these deaths, falls turn out to be the most common cause of nonfatal injury for all but one age group (males fifteen to twenty-four). U.S. hospital emergency rooms report that falls are the most frequently treated nonfatal, unintentional injury, accounting for 7 million visits a year.[11]

Around the world, as in the United States, falls are the second-leading cause of accidental death, according to a World Health Organization accounting of mortality in thirty-six (mostly) developed countries. The nations with the highest rates of falling deaths are Hungary, the Czech Republic, Norway, Slovenia, and Finland, while the lowest rates are found in Albania, Mauritius, Bahamas, Argentina, and Chile. Nobody has studied what causes these differences in falling rates.

What about the situation in China, the homeland of many of Ringling Brothers' best young balance artists? Given what we know about balance and why people lose it, it's likely that people living in China have balance superior, in general, to

U.S. citizens'. This is just a hypothesis, but here's why I think it holds water. First of all, the median age is about four years lower in China than in the United States—32.7 compared to 36.5. If balance naturally declines with age, then China would have the edge here by default. But this isn't where China holds the real advantage. Nor is it the fact that far fewer Chinese are overweight or obese (although they are beginning to catch up with us). It has more to do with activity, the types of daily physical movement that are known to challenge and tone the balance system.

In the last thirty years or so, most authorities agree, the United States, along with the rest of the Western world, has witnessed an unprecedented trend toward a sedentary life-style. It's safe to say that we now move less than we ever have during our 100,000-year history as a species. Frank Forencich, an exercise expert and human biologist, reports that a University of California researcher calculated that the average American now walks about 350 yards per day.[12] Forencich contrasts that with the several miles that modern hunter-gatherers walk daily, which is probably a fair indication of the mileage humans walked for most of our history. "We see that we are now walking significantly less than 10 percent of the historical average for our species," he writes. "This is a radical, drastic change in a fundamental behavior. If we were to take any other animal species and restrict its normal level of movement by 90 percent, it's a sure bet that we'd see some serious health consequences."[13]

Inactivity has been shown to cause a legion of illnesses, including type 2 diabetes and cardiovascular disease, and is responsible for at least 200,000 deaths a year in the United States alone. Labor-saving devices, from leaf blowers to moving walkways in airports, are part of the problem, and the one that has saved us the most labor, in a day-to-day context, is probably the automobile. By the late 1990s, 86 percent of all "person trips" in the United States were by car or truck, and well over 60 percent of adults owned vehicles. In China, however, only 16 percent of the population owns a car, with 66 percent relying on bicycles and other nonmotorized transport.[14] Having to walk or ride a bike to work, to the market, to a friend's house, ensures that most people will be active for at least some part of every day. And because balance can be maintained or improved only through physical movement, it stands to reason that the Chinese would be, on the whole, better balanced than Americans.

Transportation is just one area where the Chinese are more active than Americans. Another example is the widespread participation in a daily exercise program called tai chi (pronounced *tie jee*). At dawn across the nation, nearly every city park fills with practitioners of the fluid, slow-motion routine, part dance, part "moving meditation," as it is called. Though tai chi derives from a martial art several hundred years old, complete with full-speed parries and thrusts, the Chinese government in the 1950s transformed it into a gentle exercise for the masses to encourage healthy aging. It caught on with spectacular success—one authority believes that more than 200 million Chinese participate regularly[15]—and has since spread throughout the world. Though people of all ages do tai chi, it is designed for the middle-aged, and reportedly most people begin after fifty, with many continuing until their seventies and eighties.

A Westerner viewing tai chi for the first time might ask, "How can this be considered exercise?" After all, there are no vigorous marine corps boot camp–like drills, no jumping jacks or push-ups, no sweat-lathered bodies. The movements are done so slowly and methodically that it looks as though they couldn't possibly have a beneficial effect on the body. But there's more here than meets the eye. Ignoring the spiritual side of tai chi, which like other Taoist healing arts involves the flow of life energy through various meridians of the body, the physical movements themselves are said to improve one's sexual vigor, promote restful sleep, lower blood pressure, and modify body alignment.[16] Evidence of these benefits is mostly anecdotal, but tai chi's improvement of physical balance has been corroborated by several Western studies in the last decade. Most have used elderly subjects, both "robust" individuals and those "moving toward frailty." At the end of the study periods, subjects who did tai chi reduced their frequency of falls by 20 to 55 percent compared to the control groups.

One of the leading American researchers on this subject is Steven Wolf, a professor and director of research in the department of rehabilitation medicine at Atlanta's Emory University School of Medicine. At the end of one four-month study of tai chi, Wolf asked an eighty-six-year-old man what he could do after the study that he couldn't do before. "The man smiled at me," Wolf said, "then lifted up one foot, bent over, and took off his loafer while balancing on his other leg. Then still balancing on one leg, [he] stood back up, bent over again, and put his loafer back on....He couldn't do this before studying tai chi."[17]

And therein lies much of the power of the exercise. Many movements are performed on one leg, which challenges and improves balance in itself. Bruce Frantzis, one of the best-known American-born teachers of tai chi, describes its balance-promoting effects in his *Big Book of Tai Chi*: "Tai chi makes you better able to feel internal landmarks within your body such as your arms, spine, and hips, that tell you if your balance is solid or not. Although many exercises focus your awareness on your upper body, tai chi makes you equally aware of your lower body, which determines your balance. For many, age causes the body to lose its ability to feel, and with it goes physical balance. Tai chi increases your capacity to feel how your foot touches the floor and with what kind of pressure, which is essential to balance."[18] He says it also improves balance by the act of shifting weight from one leg to the other, which strengthens leg muscles, and by continual waist-turning movements that increase muscle strength in the hips and spine.

The beauty of tai chi, it seemed to me after taking a two-month beginner course, is that you can adapt it to match your level of fitness. The frail elderly can do basic moves without taxing themselves too much, while more athletic folks can lower their stances, for instance, for greater muscular, balance, and aerobic challenges. But although tai chi is being offered in more and more health clubs and senior centers in the United States, it's doubtful it will ever attain the popularity it has in China. Not only are Chinese practitioners considered good comrades in the Communist cause when they do tai chi, but there's a spiritual dimension to it that taps into other deeply cultural impulses.

What Frantzis says about the feet losing their "ability to feel" as we age jibes with what researchers have discovered recently about proprioceptive loss in older people. The decline varies according to body location. The hands, for instance, retain most of their sensitivity late into old age, while the feet are particularly vulnerable. Their sensory alertness holds steady until about age forty, then goes down 20 percent in the next decade, and by a staggering 75 percent by age eighty. Gerontologists say that because of this sensory loss, thick-soled shoes, like running shoes, are probably detrimental to balance in the elderly. Wearing this type of shoe is like trying to play the piano with gloves on; they desensitize the feet to proprioceptive input. If feet are already lacking such feeling because of age, then wearing a thick-soled shoe could reduce sensitivity enough to cause a stumble.

That hypothesis was tested in the early 1990s for the first time by Steven Robbins at Montreal's McGill University.[19] He selected a group of elderly (sixty-five to eighty-three) and younger (twenty-two to thirty-seven) subjects, who stood on various angled blocks, either barefoot or while wearing the same model of a thick-soled running shoe. They were asked to estimate the angle of the slope they stood on. Barefoot, the older subjects had a mean estimation error that was 162 percent greater than that of the younger group (5.13 degrees versus 1.96), which demonstrates the disparity in proprioceptive function between the two age groups. But wearing shoes caused even larger errors in both groups: the oldsters were off in their estimates by 6.58 degrees, an increase of 28 percent over their barefoot error; and the youngsters miscalculated by 3.97 degrees, 103 percent more than their mean error when unshod. Robbins concluded by recommending footwear with thin hard soles for elderly people who have stability problems, which goes along with what other gerontologists have been advising. And he hinted mysteriously that it might be possible to design shoes for all age groups that provided improved "foot position awareness," a subject we'll come back to later.

Athletic shoes, industry reports show, account for roughly a third of the American market, and it seems as though everyone, from the moment they learn to walk, wears athletic shoes, at least for some part of the day. Not a good sign for balance. Do the balance-savvy Chinese know better than Americans when it comes to footwear? As many alert consumers know, China produces roughly a third of the world's shoes. (Guangdong Province alone has more than 5,000 shoe companies that manufacture 3 billion pairs of shoes per year, accounting for 30 percent of the world's production and 50 percent of China's total production.)[20] But does that mean the Chinese wear the shoes as well as make them? Though most Chinese people don't have the income to afford $100 Nike Air Jordans, despite the fact nearly 100,000 of them reportedly work for Nike, they probably do buy and wear cheaper knockoffs. So if there was a time (perhaps before the 1980s when American and German shoe companies began producing goods in China) when Chinese people wore more traditional, thinner-soled shoes, that time has probably gone. Citizens of both nations now wear the same sort of proprioception-muddling shoes, so neither country has the advantage.

You could say the Chinese have a fetish for feet, not just because they produce such a huge volume of shoes, but because of the widespread use of the healing art known as reflexology. Practiced for centuries in China and elsewhere in Asia, reflexology involves the application of pressure and movement to the feet and hands, sections of which correspond to various organs, glands, nerves, and muscles of the body. A trained reflexologist massages and manipulates the hands and bottoms of the feet, allegedly relieving stress in afflicted parts of the body. The Chinese are so taken with this therapy that many parks and other public areas feature reflexology paths, made of concrete embedded with cobblestones of various sizes. Tens if not hundreds of thousands of people go to these parks daily, remove their shoes, and walk barefoot along the paths, the stones stimulating the bottoms of their feet in a sort of self-applied acupressure treatment.

As with tai chi, there is plenty of anecdotal evidence for reflexology's value, including claims of improvement in both physical and mental well-being, sleep, and pain relief. And again, as with tai chi, no rigorous scientific testing of cobblestone-path walking had ever been done until an American research lab decided to study it recently.[21] A hundred physically inactive adults, ranging in age from sixty to ninety-two, were selected from among residents of Eugene, Oregon. They were given tests to determine their physical strength, balance, and blood pressure. Then one group was trained to walk on special cobblestone mats imported from China. Each mat was about six feet long and eighteen inches wide, with randomly placed artificial "stones" attached to it. The control group walked on a smooth surface for the same distance and duration. After sixteen weeks, those who walked on cobblestones were "found to have improved significantly more on two balance measures ["functional reach," one's ability to reach forward in a standing position with feet planted, and "standing balance," a measure of the time a subject can maintain three different types of stances, each progressively more difficult]...than those in the conventional walking group."

Naturally, I was intrigued by this result, and set out to discover how it might occur. Knowing that a reflexology path had recently been built at a local university, I drove there for a test walk. I found the path, one of just a few in North America, on the far edge of an herb garden, hidden by vegetation from the hubbub of students. A bench was located on either end of the sixty-foot path for people to remove their shoes and socks. A small sign affixed to the back of it read: BASED ON ANCIENT CHINESE WISDOM, WALKING ON THE PATH BAREFOOT MASSAGES AND STIMULATES TRIGGER POINTS CONNECTED TO VARIOUS MERIDIANS OF THE BODY. THE PRESSURE OF THE STONES ON THE FEET COMBINED WITH GRAVITY PROVIDES A THERAPEUTIC EXERCISE TO STIMULATE WELL-BEING AND HEALTH. As I sat on the bench, I thought about those "meridians of the body" and how similar reflexology sounded to tai chi (and yoga and many other Chinese healing arts as well).

If you asked a Chinese person about how he thought these arts worked, he would probably embark on a discussion of Chi, or life energy, and how it flows or doesn't flow through the body. Believing in what they cannot see, the Chinese have faith that prescribed movement mysteriously heals the body, and perhaps this trust is the source of its power. American research on tai chi, and now cobblestone-mat walking, has looked only at the physical action the body undertakes in performing these practices.

As I began my walk down the cobbled path, I quickly discovered that some of the stones were formidable. Though they were river rocks, smoothed by moving water, some were quite thin and narrow, jutting up from the concrete substrate as much as three or four inches. Stepping on them was about as pleasurable as walking across asphalt barefoot on a hot summer day. The trick was to learn which rocks could be stepped on without eliciting pain, and I began to walk as if picking my way through a minefield. Each footfall was calculated. I applied weight slowly, waiting to see how my soles would react, splaying my arms to keep my balance as I tested the cobbles. If it seemed okay, then I went ahead and took a step. About two thirds of the way through, I thought about aborting. The bottoms of my feet were aching, and I recalled the warning sign declaring that walkers might experience "discomfort, pain, and soreness" and should not feel the need to walk the entire path the first time. Still, I kept on. After finishing, one particular area of my right foot felt a little bruised, and a brochure's reflexology chart showed that this indicated a problem with my lungs. Never having had any trouble with these organs, and leaning toward a literal view of the world, I took the pain to mean that a pesky cobble had simply abused this part of my foot.

I chose not to double back on the snaking path, but sat for a while on a bench, enjoying the tingling sensation on my soles, as if I'd just had a thorough massage. And I now understood how cobblestone walking could help one's balance. Navigating the path, I felt sort of like a funambulist, always challenged to maintain my center of gravity over my feet. It seemed as though my feet were "naive," perhaps from always being enclosed in cushioned shoes, insulated from the texture and feel of surfaces. They had a hard time dealing with the hard irregularity of cobblestones, though I'm sure they would get "smarter" with practice. The Oregon researchers wrote that cobblestone walkers "required more attention to maintaining balance while standing and transferring weight during locomotion. Although not measured in this study, the exercise may have affected proprioceptive and kinesthetic awareness, making participants more conscious of their postural limitations and requiring them to make appropriate postural adjustments."[22]

After the walk, I got to thinking about what Forencich had said about the benefits of walking on rough, uneven terrain. He believes that living in a world of hard, smooth surfaces doesn't provide much stimulation to our nervous systems, which therefore become less efficient. He encourages people to go out and hike on trails that undulate and dip and angle, strewn with rocks or boulders. Thin-soled trail shoes, rather than heavy hiking boots, would be preferable to wear because they would allow more sensory perception on the bottoms of the feet. "Every little rock, every uneven root, every slippery patch of moss...boosts the detection of tactile signals," he writes. "Rough terrain wakes up the sensory nervous system and makes your body smarter."[23] Although I'm sure he didn't have cobblestone paths in mind, he could just as easily have been describing their effect.

It should be obvious by now that the way a nation spends its leisure time has a lot to do with how well its citizens maintain their balance. With that in mind, here's one final observation about the homeland of tai chi and reflexology. China's number one sport is not a spectator sport, as it is in the United States (where the Big Three are, of course, football, baseball, and basketball), but one in which hundreds of millions of people, old and young, men and women, participate. You may have guessed it: table tennis. Among those who consider it a sport, not a basement pastime, table tennis requires superb agility, aerobic stamina, leg strength—and great balance. Though it originated in England, Ping-Pong (its trade name) was embraced by Chinese Communist officials in the early 1950s, much as tai chi was. They considered it not only a sport that could be played anywhere, by practically anyone, but one whose physical demands were well suited to the average Chinese build: short of stature, quick of reflex. The Chinese took the sport very, very seriously, and within a short time Ping-Pong schools and clubs sprang up nationwide, and millions of people began to play. Today, China wins a vast majority of world-level tournaments, including numerous victories at the Olympics (table tennis was included in Olympic competition in 1996).

I know an American, Gene Treneer, who began playing the game seriously when he was fifty. It consumed him, and gradually he became good enough to play competitively. Today, at age eighty-four, he plays four or five times a week and takes a weekly lesson from the top professional player in the United States, an émigré from China named Yi Yong Fan. Gene has never fallen, on the table tennis court or elsewhere, and he attributes most of his agility and balance skills to the sport. He likes to joke that he's rated number eight in the country in his age group, mainly because there just aren't that many competitive players over eighty.

I've watched Gene hold his own in casual match play with several Chinese players. One was a defensive specialist, perhaps a decade younger than Gene, who preferred playing barefoot—for greater proprioceptive sensitivity perhaps? He stood ten feet back from the table, running down every aggressive topspin shot Gene hit, returning them with wicked underspin. With nary a stumble, trip, or falter, both men adroitly moved to the ball, smacking it back and forth in a graceful dance, sweat dripping off their faces. You could tell they were loving every second.

It turns out that some experts believe racket sports are one of the better overall exercises for improving balance. Karen Perz, the physical therapist from chapter 2 who specializes in vestibular rehabilitation, offered some insights as to why they're so effective: "You're exercising a lot of aspects of the [balance] system. You're watching something that's moving because you're chasing the ball with your eyes, you're turning your head, you're moving your feet. And so you're getting a good aerobic workout and a lot of stimulation with head movement. You're getting a lot of visual stimulation, too, plus somatosensory stimulation, and you're developing motor skills at the same time."

But racket sports are by no means the only avenue to good balance. Another effective "therapy" is dancing. "Ballroom dancing, African dancing, folk dancing—it doesn't seem to matter," says Mary Tinetti, chief of geriatrics at the Yale University School of Medicine. "They all appear to help people know how to respond when their balance is challenged."

As surprising as it may sound, Karen Perz is also a big advocate of gardening as a balance-enhancing activity. "For someone who enjoys gardening," she says, "it's a very good exercise for balance because of the combination of bending and straightening and walking on uneven ground." Gardening is also one of the best "real-life" opportunities to sneak in exercises to develop lower-body strength, agility, and balance, says Forencich. When you squat down to pull weeds, you can pay attention to your form and lower yourself slowly, and perhaps throw in a few more reps for maximum benefit. Forencich is a big fan of finding such opportunities to challenge the balance system because it's an easy way to incorporate useful, "functional" exercise into everyday life.

One of the hottest buzz phrases in the fitness profession these days, *functional exercise* is simply movement that mimics real-world activities. Of course, the specifics depend on what activities you want to be fit to do. People who watch a lot of television might just want to be strong enough to raise and lower themselves from the couch without falling, but most people would like to be able to do things like play tennis, play with their grandkids, hike, ski, or walk a few miles around the neighborhood. So instead of doing reps on a weight machine, which works just a few muscles in isolation and makes you better at performing only that exercise, you would, say, throw a weighted medicine ball up in the air and catch it. In order to accomplish this feat effectively, you need to use not just your arms but your legs and torso as well, so you integrate the action of several muscle groups, teaching them to work together and enabling you to perform a wide range of activities more easily, from walking to throwing a Frisbee to taking out the garbage. You don't need specialized equipment to do these sorts of practical exercises. A laundry basket will suffice. "Laundry is something we all have to do," says Forencich. "Instead of bending over at the waist [to pick up the basket], go to squatting movements or lunging movements. Pick up your socks with a squat or a lunge. Or if you want to work in some balance, you can pick up your socks on one foot. You can make a little mini–athletic thing out of it. If you can imagine how Jackie Chan was going to do his own laundry, that's probably how he would do it."[24] Doing squats and lunges strengthens your quadriceps, an important set of muscles for maintaining equilibrium, and challenges your balance system as well. Among older people, quad weakness is a major contributor to falls, which often occur while squatting, for instance, into a chair or bathtub.

Achieving better balance and stability are two of the primary goals of functional exercise. Most fitness and physiology experts will tell you that balance is a skill that can be learned or unlearned, just as muscles grow stronger with training or shrink with disuse. As with muscles, the balance system needs to be challenged in order to improve. As Forencich observes, "The body adapts with exquisite precision to the stresses that are imposed upon it....The language of the body is challenge and response." In the last decade manufacturers have created an assortment of balance-challenging equipment to meet the rising popularity of balance training in fitness clubs, physical therapy clinics, and home gyms. The idea behind all of them is the same: to create an unstable platform for the body, forcing the involvement of so-called stabilizer muscles of the spine and hips, small muscles difficult to target with conventional exercise. These muscles are critical for controlling posture, as are the neuronal connections between them and the central nervous system.

Over the years, I've tested many balance devices, to see how fun, challenging, and effective they were. One, the venerable Bongo Board, was popular forty years ago among skiers. It's a rectangle of plywood that rolls back and forth on top of a cylinder. Riding one is a precarious venture for beginners, who can easily be tossed onto the floor. I managed to learn the skill fairly quickly, without injury, and though it was fun for a while I grew tired of it as my ability increased. Other balance boards are less tippy and therefore suitable for people who are cautious or older. One, called the Belgau Balance Board, is easily adjustable; when your ability increases, you can swivel the twin rockers underneath the platform to bump up the challenge. Another good choice is the wobble board, a circular-shaped foot platform perched atop a fixed half-dome base. Both the Belgau Board and wobble boards are great for doing dumbbell exercises, squats, or simple balancing.

Another superb balance-training device is the oversized plastic ball known by various names: gym ball, Swiss ball, Physioball, exercise ball. Developed in Switzerland decades ago for injury rehabilitation, it is now as common as sweat in gyms and physical therapy centers. The ball's innate instability makes it ideal for scores of balance-challenging exercises, ranging from crunches to back bridges. While older people can choose relatively safe movements, the difficulty level can quickly approach the stratosphere; some athletes have been known to kneel or stand on a ball as they work out with weights, and circus performers often are trained to "walk" balls along for great distances.

But though these tools (and toys) are touted to improve balance, and indeed they can, they have their drawbacks. First of all, you can fall off of them, so they can actually contribute to the problem they're supposed to prevent. Some are just too challenging for older or less active people, who shouldn't use them without the guidance of a wise physical therapist or trainer. Also, learning balance on such a device doesn't necessarily confer good balance in real-world activities. They challenge balance in a limited way. Just because you can balance on a Bongo Board for minutes at a time doesn't mean you'll be a better snowboarder, for instance. And for people who don't like gadgets, such devices may be superfluous. One of the most effective and portable balance gadgets is attached permanently to the human torso: your legs.

As both a gauge of balance ability and a way to improve the skill, one-legged standing exercises are among the most versatile and effective challenges. Doctors and physical therapists often employ them as a diagnostic tool to assess a patient's equilibrium. Though one-legged balancing sounds easy (heck, flamingos and many other shorebirds stand on one leg for hours at a time, as a way to conserve body heat and perhaps to make them less conspicuous to prey), it can be difficult for older people or those with balance maladies. Physical therapists compare a patient's performance to a standard "normal" range, which gives a good indication of how balance diminishes with age. For people twenty to forty-nine years old, it's twenty-four to twenty-eight seconds; fifty to fifty-nine, twenty-one seconds; sixty to sixty-nine, ten seconds; seventy to seventy-nine, four seconds. And most people over eighty can't do it at all. As you stand on one leg, the muscles and ligaments around your ankle are firing like crazy as they attempt to keep the body upright. If you can do this while keeping your body still, you're relying mostly on proprioception from the feet, ankles, and legs to maintain balance. Input from vision is secondary; closing your eyes makes one-legged standing more difficult but not impossible. The vestibular system kicks in only if you struggle to stay still and need to make large hip and torso movements to maintain the position. People with bilateral vestibular dysfunction, such as Cheryl or Robin in chapter 2, often have trouble with one-legged standing and activities like walking on a narrow beam.

A surprisingly wide range of balancing exercises can be performed on one leg (see the appendix). One reason they're so effective is that humans, during locomotion, can more accurately be called monopeds than bipeds. "In normal human gait, stance phase (when a foot is on the ground) and swing phase (when a foot is moving through space) alternate in a complementary fashion," says Forencich in *Play as if Your*

Life Depends on It. "All of the time that you are swinging one leg forward, you are in fact standing on one foot." Researchers have determined that about 80 percent of walking, and 100 percent of running, is spent on one leg at a time. "By building balance and useful strength in one leg at a time," Forencich contends, "stance phase becomes more stable and gait improves." The other advantage of one-legged exercises is that they accomplish two things at once, building lower-body strength and fine-tuning balance. Plus, they can be done anytime: while brushing your teeth, waiting for the bus, watching television. Instead of sitting down to put on socks or take off pants, fit and healthy people of all ages can do it while standing on one leg, slowly and in control; likewise, after a shower, dry your calves and feet by raising one knee up to your chest, balancing on the other leg, assuming a flamingoesque posture. (Of course, if these moves are too challenging it's best not to risk a fall, especially not *in* the bathtub, but you get the idea. A one-legged stance can mean simply shifting most of your weight to one leg while keeping both planted on terra firma, as in tai chi.)

One-legged stances are a proven means to improve balance. Three variations were used in a study at Indiana University in 2005, designed to see whether a simple, inexpensive, home-based program could bring about positive results.[25] A group of healthy fifty-five- to sixty-year-olds practiced the postures fifteen minutes a day, four days a week, for six weeks. At the end of the study, researchers found that the subjects' "sway patterns"—the subtle, almost undetectable movements the body makes, front to back and side to side, as it reacts and adjusts to the destabilizing effects of gravity—had improved (made the body more stable) by 16 percent. "We took the balance system, and instead of being more sluggish, it became more flexible and adaptive," said David Koceja, a professor in the university's department of kinesiology and an expert on balance and posture control in older adults.

These sway patterns change with age, a result of degeneration in vestibular, proprioceptive, and musculoskeletal systems. After age sixty, the circumference of the elliptical-shaped pattern becomes larger, as if the body were a tree whose roots have lost their grip on the soil, making it more vulnerable to the destabilizing forces of wind, ice, and snow. Scientists believe this widening arc shrinks what's called the "limit of stability" beyond which a fall is likely to occur. The body's center of gravity moves more rapidly in response to the greater sway, and "the momentum of the body acts as an additional destabilizing force," according to Claude Hobeika, a physician and director of the Ear Medical Center in Cincinnati.[26]

Older people who exercise regularly or participate in sports, one study showed, have a smaller sway pattern, and thus better balance, than those who aren't active. One of the most interesting aspects of the study, published in the *British Journal of Sports Medicine* in 1999, was that there is almost no difference in the sway patterns of elders who have exercised all their lives and those who began after retirement, indicating that it's never too late to start a strength- and balance-training program.[27]

One area of balance training that's rarely addressed, even in the more progressive health clubs, is vision. As I've pointed out many times, vision is critical for balance in many circumstances. So it makes sense that by improving vision you can also enhance balance. But, you might ask, besides using corrective lenses or having surgery, how is it possible to improve your ability to see? It turns out that visual acuity—how accurately you can read the letters on an eye chart from a certain distance—is just one aspect of vision. Your eyes are doing lots of other things, too, such as tracking moving objects, changing focus from near to far and back again, and perceiving depth. Optometrists who specialize in what's termed "vision therapy"—exercises that challenge the eyes in specific ways—claim that these other components of vision can be improved through training. Along with improved vision, they say, people who do vision therapy can achieve better balance as well. (See the appendix, page 272, "Walking the plank," for an example.)

Though few studies have attempted to prove this assertion, at least one vision therapist, Burton Worrell, has done research that appears to bolster the idea. He studied a group of twenty-eight college baseball players to determine what, if any, visual traits distinguished the better hitters. He and a team of doctors put the players through a battery of tests, but they could find little correlation between good hitting and eye movement skills or sharpness of vision. They did, however, discover six skills, two related to balance, that seemed to correlate. "Players from the lower half of the batting average list," Worrell says, "fell off a balance board [a length of two-by-four lumber, laid flat on the ground] with their eyes closed. They also fell off the plank while focusing on something moving in front of them."[28] So Worrell devised a series of exercises, including walking the plank, designed to challenge their vision and balance systems. After twelve weeks of performing these exercises, players had improved their batting averages by an astonishing forty-three points. While the researchers aren't sure about the mechanisms that caused the improvement, Worrell offered a possible explanation. Most batters hit most effectively when they're in a balanced position. The pitcher's goal, Worrell said, is to throw the ball

to areas around the plate—low, high, inside, outside—that
cause a batter, if he swings at it, to be off balance. Because
of the connection between the vestibular system and vision,
Worrell says, when a batter is unbalanced, his vision is often
impaired. "He becomes nonvisual for a fraction of a second.
Sometimes the batter won't see a pitch because he was thrown
off balance by the pitcher." Performing vision-therapy exercises, Worrell and other advocates claim, trains the eyes to
stay focused on objects even when the body is not in perfect
equilibrium, improving athletic ability in a wide range of
sports, as well as daily living activities.

There's no single reason why Americans are now having more
problems with balance, and falling, than ever before. But with
the rising epidemic of obesity, the growing lack of regular activity, the greater number of medications people take, and the
aging of the population, it's no great surprise that so many
people are unbalanced. What's most disconcerting, though,
is what's going to happen in the next few decades, as baby
boomers begin teetering into old age. By 2040, the number
of people sixty-five or older is projected to more than double
from its current level, to 77 million. That means, at the current rate, 25 million people will fall every year, resulting in
8 million injuries and 25,000 deaths. By then, however, the
rates of both falling and death by falling will probably have
risen to even more disastrous levels, in keeping with the general upward trend. (From 1970 to 1995, the rate of falling
among the elderly increased 124 percent, and the death rate
from falling rose 80 percent.)

But the good news is that while balance does deteriorate with age, experts claim that up to half of all falls are
preventable. Exercise is part of the solution, as recent stud-

ies have shown. Older adults can maintain their equilibrium by strengthening key muscles and stimulating the balance system. One such investigation was conducted in 1996 by the department of neurology at the University of Connecticut School of Medicine in Hartford. It showed that after just three months of doing balance exercises, a group of older men and women achieved a level of balance similar to that of people three to ten years younger.[29] The hard part, of course, is persuading people to get up off the couch and get active.

Anyone with a balance problem, and that includes most adults over sixty, can take a tip from Karen Perz, who often tells her balance-impaired patients to keep a few simple things in mind when moving. Watch where you're going (how many times did you hear that growing up?). Do one thing at a time (research has shown recently that human performance degrades markedly as the number of activities increases). And give your balance the attention it deserves. After musing on her words, I thought they could apply to just about anybody because we are all, at one time or another, in one form or another, "challenged" to stay upright and balanced. A friend of mine, who has no obvious balance problems, sued the City of Seattle because he didn't heed these caveats. Walking in a city park with a companion, deeply absorbed in conversation, he stepped into an unmarked hole and tore ligaments in his calf.

As imbalance becomes more common in the coming decades, the general public will be called upon to assist people with balance problems. But how do you do that effectively? Karen Perz offers her wisdom: "When someone is off balance, it's better for them to hold on to you, rather than for you to hold on to them. If you're holding them, then what happens is they're having to react to you, as well as to whatever the trouble is. And when they start to get off balance, they'll begin to make their own correction, and you've seen that they're off balance so you're correcting as well. But the part that you're correcting, they don't know what's coming, so you can inadvertently pull them off balance or make it more difficult for them. And most people don't know that you should offer your elbow, not your hand, because it's more stable. If they can hold on to your elbow, they're controlling the amount of help that they get and it's predictable. They can adjust to what they need, and they know what they need probably better than you do."

For people who are still young enough to have sufficient natural balance, the take-home message is pretty clear. If you want to maintain your balance right through to the end, you've got to challenge and stimulate it with balance-specific sports and exercises. The younger you start, the easier it will be to make it a lifetime habit and the better balance you'll have throughout your life, though it's possible to make big gains no matter when you begin. In the case of elite athletes, like circus performers, those balance challenges were first met in childhood and mastered after a long period of adaptation. If the challenge continues regularly throughout one's lifetime, age doesn't seem to have much power to degrade the adaptation, as evidenced by the longevity of most of the world's supreme high-wire walkers. While we might think of them as freakish in their abilities, almost superhuman, usually the only difference between them and the average person is the amount of time they've spent practicing. For we all have roughly the same innate capacities. It's only been in the last century that we've fallen from the ancestral pattern of regular movement and activity. Descended from a long line of arboreal acrobats, we are all, at our cores, wire walkers and hand balancers.

Chapter Nine

The Cognitive Connection

Everybody knew that low gravity made you dumb — the "space stupids," they called it. Mikhail was well aware of the importance of concentrating on the chores necessary to keep himself alive.

— FROM THE SCIENCE-FICTION NOVEL *Sunstorm*, BY ARTHUR C. CLARKE AND STEPHEN BAXTER

During missions in space over the past forty years, a number of astronauts have reported a strange malady, a sort of mental fog that envelops their brains and dulls their reasoning. It's so common, in fact, that there's a name for it: "space stupids." No one is sure what causes it, few have studied it, but a number of NASA researchers believe it has a vestibular origin.

"When they first get up there," says Dan Merfeld, the head of the Harvard vestibular lab and a former NASA researcher, astronauts "simply are not working with a full deck. Some will admit this and some won't. They're not thinking nearly as clearly as they were on the ground." The cause isn't absolutely evident. Stress might be a factor. But there's also a strong possibility that it's due to some sort of link between the vestibular system and cognition—mental tasks such as perception, learning, memory, and reasoning. Because this hypothesis hasn't been proven, Merfeld withholds comment on its soundness.

Don Parker, the vestibular researcher at the University of Washington who has done a lot of work for NASA, also sees a possible connection between the two. "I think the major interest for me in the vestibular system is that it's fundamentally a spatial orientation system. It lets me know how I'm oriented and how I'm moving in space. You can't act unless you're oriented. It follows that the vestibular apparatus, as a contributor to the spatial orientation system, is fundamental for behavior. Appropriate action [of the vestibular system] is very important for behavior. When you screw it up, then you're stupid." What's screwing it up, of course, is the sudden dramatic reduction in gravity in outer space, and perhaps the unusual motions the body encounters during space travel.

The relationship between cognition and the vestibular system, depending on whom you ask, seems obvious, inconceivable, or tantalizingly possible. Those who demand large-scale, controlled, peer-reviewed, double-blind studies aren't convinced, for these kinds of studies simply haven't been done. For such skeptics, the vestibular system's sole function is to stabilize posture and eye movement. Yet a number of recent experiments on the brain, combined with anecdotal evidence from fields as diverse as medicine, space flight, and education, hint that the body's dedicated balance organ may play a role in learning, judging, knowing, and thinking.

Though the closest she's been to outer space is the cruising altitude of a commercial jetliner, Cheryl Schiltz, the woman who, after being given a common antibiotic to treat a post-surgical infection, suddenly lost most of her vestibular function, has experienced something alarmingly similar to an astronaut's space stupids. Her cognitive impairments came immediately after her diagnosis of BVD (bilateral vestibular dysfunction), when processing information of all sorts suddenly became a challenge. "Conversations, you can do one-on-one really well, but if I'm in a room with a lot of noise going on and numerous conversations going on, it's a nightmare," she says. "It's too much coming in. I can't deal with it. When you have that sensory loss, what happens is that it messes you up in such a way that cognitively you begin to have a lot of problems. Short-term memory, just trying to figure things out, adding two and two....Reading is very difficult. Dyslexia for sure. You turn everything around. Aphasia [the loss of a previously held ability to speak or understand language] is common. I could literally taste and I could see the word in my head, but I could not say it. I'd have to pick up an item and show it to someone, and ask, 'What the hell is this? Tell me the word, I can't get it out.'"

If these were the experiences of a single person with BVD, then her bizarre symptoms might easily be shrugged off. But it turns out that Cheryl's story is common among people who suffer from this and other disorders of the vestibular apparatus. Cheryl told me about a speech she'd read on the Web by a Portland, Oregon, psychiatrist about the cognitive aspects of vestibular disorders. She said it touched on every symptom she experienced, so much so that she could have written it herself.

The speech, which Dr. Kenneth Erickson presented several years ago at a Vestibular Disorders Association conference, outlined cognitive impairments he'd observed in his vestibular patients.[1] Many exhibited a diminished ability to multitask, to monitor more than one mental process at a time. Another common complaint was difficulty in handling sequential tasks such as the proper ordering of words and syllables in speech. Mental stamina, memory, and something he called "channel capacity"—the flow rate of new information a person is able to process—were all reduced in a majority of Erickson's vestibular patients.

In his talk, Erickson anticipated arguments psychologists might raise about other likely causes of this strange pattern of cognitive faults. Perhaps it didn't have anything to do with the vestibular system. Perhaps it arose from pain, fatigue, or depression, three common complaints from people with severe vestibular disorders. Erickson acknowledged that all three would indeed be likely to affect a vestibular patient's performance on routine psychological tests. But with the right kind of testing, a distinction can be drawn: "If we test people with a lot of pain or depression or fatigue, they will do badly on a variety of attention and concentration tests. On those tests, however, vestibular patients may do pretty well...Obviously common sense leads us to explore this further. We can only conclude that this kind of malfunction seems highly specific to most vestibular patients."

When I spoke to Dr. Erickson, he apologetically explained that the data he had used in his lecture was from a study he conducted in the early 1990s, which he never published. He said that his study had looked at thirty patients with vestibular disorders, comparing them to thirty age-matched "normals." He had also drawn from his experience evaluating about fifty other vestibular patients before then. Since the study, he has also treated "three or four hundred" more, many referred to him by doctors at the Legacy Good Samaritan Hospital and Medical Center in Portland, which has one of the top vestibular research centers in the country. Most displayed some sort of cognitive problem, he said, though he never followed up on his initial study nor sought to pinpoint the causes.

Explaining why he thinks cognitive problems exist in vestibular patients, he invoked an argument he credited to a Portland neurologist, Dr. Robert Grimm. "It appears that what our brain needs to function well is unchanging, accurate, balance-and-gravity, up-down type of information. And while the semicircular canals of the inner ear are the starting point, the signals from there go to the brain stem and up into the brain in many different areas. The parietal lobe, which is where we form many of our mental maps of the world, is a logical place to be looking.... There are over thirty different what we call somatotopic maps in the brain, which means that a group of neurons have to send their signals to another group of neurons, and there has to be a point-for-point accuracy, when the maps are sent from one point to another. And these maps begin to confound, alter, and distort when poor or shifting balance signals keep coming in. It's just a theory, there's no way to measure it on a brain scan, but it's a reasonable one."

I was struck by how similar this argument sounded to Don Parker's explanation of space stupids. Unfortunately, without a rigorous, peer-reviewed analysis of Erickson's work, it's difficult to judge how strongly to credit the theory. But Erickson can at least make some sense of things for patients who complain of cognitive difficulties. "With a patient I use the analogy of older encyclopedias, ten to twenty years ago," Erickson continued. "If you had a map of China, with plastic overlays of the cities and the rivers and the elevations, you need them to overlay accurately to make sense out of the picture. But what if those maps were swirling around, moving a bit here and there, distorting? You would have very confused information. You would have oases of clarity, but there would be a lot of areas that were distorted, and you would be very uncertain of yourself.

"You or I could talk to a vestibularly affected person for hours, and say, 'You make sense, you can keep a logical sequence going, there's nothing wrong with you,' but later when you talk with them, they had, say, forgotten two very important things they wanted to share with the doctor. And this drives them crazy, and they won't be able to say why they forgot it. It's just gone.... And there seems to be no escaping it."

I discovered another piece of this puzzle in the oddest of places. In my search for balance-enhancing exercises and devices, I found on eBay a vintage balance board like the one mentioned in the last chapter. I scanned several pictures of the plywood platform perched on twin adjustable rockers and was intrigued by a close-up of a paragraph of text printed on the underside of the board. It read: "The Belgau Balance Platform is a very powerful tool for developing intelligence and improving brain processing efficiency.... You will also observe significant improvement in memory, binocular vision and visual acuity.... Balance stimulation can speed up, organize, and improve brain processing functions." Excited by this chance find, I put a bid in on the board but lost out. However, to my surprise, when I Googled "Belgau balance board," up popped a Web site for a company called Balametrics, started in 1982 by Frank Belgau. Paging through the site, I saw that Belgau was still manufacturing balance boards that looked almost exactly like the one I had seen on eBay. Even more amazing, the company was located in Port Angeles, an old lumber town on the Strait of Juan de Fuca about seventy-five miles northwest of my home. I arranged to visit Belgau.

Reserved yet friendly, he looked a little like a diminutive version of Colonel Sanders, short and round-bellied, with a white goatee and mustache. His voice had the rhythm and tone of Mr. Rogers, but Southern-fried, hinting at his long residence in Florida and Texas. After exchanging pleasantries, Belgau insisted that I try some of the products he uses in his so-called Learning Breakthrough Program. First he checked my eyes using a long piece of string stretched from my nose to a chart twenty feet away. I sighted along the string, first with one eye, then the other, which told him about the ability of my eyes to converge on a target. He stated that my eyes had normal convergence. Then he recorded my voice as I read a brief, difficult passage in a history book that included a lot of French words. Afterward he had me stand on a balance board like the one I had seen on eBay. The curved rockers underneath could pivot on their axes, which changed the stability of the board. On an easy setting, I stood on the board and followed his instructions to toss a beanbag in a variety of ways, feeling a little ridiculous as I threw it from one hand to the other, varying the height of the bag, always looking directly at it, sometimes passing it around behind me. Tossing and catching. Tossing and catching. The balance board didn't seem to challenge my equilibrium at all. After about ten minutes of this, I read another passage in the same book, as Belgau recorded it. We then listened to both versions, and I thought I could detect my reading pace bump up a notch, my enunciation become a little smoother. It wasn't a big change, but it was noticeable to me. And it was very noticeable to Belgau. If you're a strong reader to start out with, as I am, then the changes are apparently more subtle.

Belgau runs people through this little exercise as a preface to any explanation of the theory behind his program, which he devised during the 1960s while teaching learning-disabled children in Texas. He said witnessing a demonstration makes a bigger impact on people than merely talking about its effects. Although the influence of balance-board activities on reading ability varies from person to person, Belgau told me, sometimes it can be profound. He inserted a tape of a ten-year-old boy reading in a painfully halting manner, stumbling on almost every other word. After his first session on a balance board, tossing a beanbag for ten minutes, the boy read again. The difference was remarkable, as if a switch had been turned on that sent more electricity through his brain. On the tape, Belgau asked him if he noticed any difference. "Yes. I don't stop as much," the boy responded. "Is it any easier to understand?" Belgau queried. "Yes," the boy said.

Okay, the million-dollar question: how can something as simple (and silly) as throwing a beanbag up and down while perched on an unstable surface cause improvements in reading, learning, or any other cognitive function? There's certainly nothing innately intuitive about this relationship, if there is one, though I recalled that two of the pioneers of vestibular rehabilitation, the British physicians Cawthorne and Cooksey (whom I introduced in chapter 2), had also used beanbag tossing as part of their therapy more than sixty years ago. I wondered if Belgau had borrowed that element from them.

The story of how Belgau came up with his approach starts back in the 1940s, when World War II had drained many career teachers out of the U.S. school system. He was a junior high school student at the time, and one of his replacement teachers was a man of advanced age named Carl T. Royer. A friend of Thomas Edison, Royer made a point of sharing some of the great inventor's thoughts and philosophies with his students, and they made a big impression on Belgau.

Edison believed that if you had a problem, you tried any so-lution. If it didn't work, then you tried something else. "Edison's observation of humanity," Belgau said, "was that when most people have a problem, they come up with a solution, and whether it works or not they're stuck with it. [Edison] also observed that the more education you have or the more authority vested in you by a government or a professional society, or the more important you are in a crowd or a group, the harder it is for you to have a bad idea and give it up. Edison said you had to practice being observant. And most people missed the things that were really important because they didn't observe carefully enough. Most significant things are subtle."

After high school, Belgau joined the air force and became a flight engineer, graduating from aircraft maintenance school. A gifted troubleshooter of aircraft power plants, both recip-rocating and jet engines, he was later promoted to instruc-tor. "The thing I learned to appreciate was understanding the basic operating principles," he observed. While still in the air force, Belgau took night classes at Sam Houston State Uni-versity, and then he did graduate work at the University of Houston. Majoring in education and psychology, he hoped to learn "the basic operating principles of human beings," he said, "but they really didn't offer that. In fact a lot of psychol-ogy things seemed pretty far out."

In the early 1960s, Belgau left the air force after receiv-ing his university education and started teaching elementary school. His first class was made up of fourteen students from fourth through sixth grades, "really bright kids who had seri-ous reading and learning problems." Along with using a con-ventional approach to teaching learning-disabled children, Belgau also taught them what he had learned in junior high about Thomas Edison. "Six weeks into the class, we were having a discussion," Belgau recalled of his students. "They decided that what we were doing wasn't working. My reaction was to fall back and hide behind my certification and my professional authority. I was ashamed. All of a sudden I came to my senses and told them we'd just experiment." That's when he began exploring novel approaches, anything to get his students to read and learn more effectively. He tried tape-recording passages from a book, speeding them up, and having the kids follow along as the tape was replayed. With what he called "auditory support," the kids were able to read faster than before.

A eureka moment occurred after thinking about an observation he'd made of many of the students. When they walked, about half didn't swing their right arms in a free and natural way. "One of my dad's professors had been a scientist who had developed the mathematics for locating earthquakes, using information from seismographs and pinpointing the location," Belgau said. "Now, the seismograph is really a pendulum. And so a part of my early culture was pendulums. Pendulums were something in our house. I looked at walking as harmony of pendulum movement. The more harmonious the movement, the better the walk looks." The idea formed in his mind that the movement of the body through space is actually a picture of the operation of the brain "in one dimension." If the pendulum moved awkwardly, Belgau inferred that the brain was off-kilter somehow. He also knew that the brain is equipped with its own pendulums, its own inertial guidance system, the vestibular apparatus, that measures head accelerations and gravity. Having taken one of the air force's first classes on inertial-gravitational systems for guided missiles, Belgau knew that pendulums were a critical component of the early versions of these complicated machines, like those designed by Sperry for ships and later airplanes in the early part of the twentieth century. Deflections of the pendulums in an inertial guidance system, as in a seismograph, indicated the direction and force of an acceleration, of a cruising submarine, for instance, or of Earth's shifting plates.

After researching the vestibular system, he reasoned that it greatly simplified the developing brain's sense of the three-dimensional world. "The vestibular sense is an inertial-gravitational sensor," Belgau said. "It senses changes in inertia, and gravity is part of that. That's the way it makes sense of the world. So the first sense that actually develops after conception and starts feeding the brain information is the vestibular processes. As the baby moves in the womb, the motor [movement] processes are matched into it. And what's important is that the motor system is based on this original inertial-gravitational system, like a pendulum...that measures both time and space. The brain uses that, it's referenced back to that." And just as the pendulum is "calibrated" to the acceleration of gravity, so is the human brain, through the vestibular system. "This went through my head for a long time before it made any sense," he said. "The acceleration of gravity almost has to be—it must be—the human brain's calibration reference.... The resolution of your balance is a clue to the resolution of all your systems."

Belgau began making and experimenting with pendulums, hanging objects from the ceiling on strings and swinging them and timing the swings. He would change the length of the string to alter the period of the swings. Then he took pendulums to school with him. He hung balls from the ceiling and created "games" in which the students would use a small stick to strike the balls and control their trajectories.

Thus, following Edison's edict of searching for any workable solution to a problem, through a significant observation of a subtle clue, Belgau worked out a hypothesis of what he understood to be a basic operating principle of the human brain. But would it work?

His classroom experiments at first elicited comments from his students such as: "Good night, Mr. Belgau, you sure come up with some crackpot ideas!" This sort of response would, in fact, dog him for decades. But what girds him now against negative criticism is the same thing that buoyed him then: kids, for the first time in their lives, began to read and learn at a much higher level than ever before, he reported. Within a few years, as he began to realize the central importance of the vestibular system to brain functioning, he added balance boards to the pendulum-and-bat activities, and later beanbag tossing. Belgau's unusual methods attracted the attention of the dean of the University of Houston's optometry department, who was so impressed that he hired Belgau to create and direct what was called the Perceptual Motor and Visual Perception Laboratory.

When I spoke with Belgau, it seemed as though his seven years at the university were a source of bitterness and disappointment. The department he created no longer exists. All he would say about what he accomplished there was that he taught students to evaluate children using his methods. Though a few faculty members supported him, most were opposed. "Some of the people at the university were just really angry with me," he remembered. "They said the only reason I had any effect was because of the placebo effect. [They believed that] if you had a kid with a reading problem, you just have to sit him down and teach him to read. You're not going to teach a kid to read by putting him on a balance board." He jumped through enough academic hoops to win tenure, despite the efforts of colleagues to bar the door. But at that point, he says, he'd had enough of university life and Houston, and decided in 1972 to move to Washington State to raise a family and start his business.

Until about five years ago, he labored in obscurity, producing and selling enough Learning Breakthrough kits to make a comfortable, modest living and raise three children with his wife, Beverly. Shunned by most school districts because he lacked evidence that his system worked, Belgau ended up selling mainly to homeschoolers, whom he reached through seminars he gave at regional conventions, and to Internet customers. He bought a motor home and traveled the country, attending conventions and giving seminars. "I would drive all night and talk all day," he said.

I asked him if it wouldn't have been easier to sell his products if he had conducted or commissioned a study showing their value. Certainly, he said, but he was too busy marketing his business and spending time with his family, and besides he didn't have enough money to launch a project like that. His answer somehow didn't ring true with me. Was he afraid his ideas would be proven unsound?

Along the way, he refined his thinking about why his system seemed to work. One man who fertilized his ideas was William Calvin, a University of Washington theoretical neurobiologist. A world-renowned expert on the human brain and evolution, and the author of a dozen books on the subject, he had devised an intriguing hypothesis about how, in the early evolution of humans, the hunting of game with small hand-thrown projectiles may have spurred the growth of the brain. Calvin argued that to hit a moving target requires an enormous amount of planning by the brain, with no time for correction. When the distance to a target doubles, the so-called launch window, the precise timing of the toss, requires the activation of sixty-four times as many neurons to ensure accuracy. Belgau took this assertion to mean that activities like beanbag tossing on an unstable surface, or directing the trajectory of a pendulum ball with a stick, would engage more and more neurons as the difficulty level increased. As more neurons were recruited, the brain would rewire, or recalibrate, itself to perform in a faster, more efficient manner, Belgau surmised.

I wondered what would happen if the balance and timing challenges were more extreme, as in, say, the case of a tightrope walker juggling five frying pans while crossing above Niagara Falls. Would there be a commensurate increase in the number of neurons enlisted? After all, one of the definitions for funambulism is mental agility, as if there were a direct association between physical and cognitive coordination. Belgau didn't think so. People with high levels of physical agility aren't necessarily nimble of mind. When I mentioned my interest in wire walkers like Karl Wallenda, Belgau seemed to have firsthand knowledge of their technique, observing that, in general, they were rigid in their comportment. It was true, I later discovered from talking to the tightwire specialist at the San Francisco School of Circus Arts, that a rigid upper body is an essential part of the wire-walking technique; otherwise, a slight shift away from the wire might cause a fall. When the balance challenge is too high, Belgau said, it can "turn off a whole lot of your systems" and hinder the recalibration process. A relaxed, easy balance challenge was preferable, to get the most out of his program. Then Belgau surprised me by saying he had grown up in Sarasota, Florida, the winter home of the Barnum and Bailey Circus. Many circus performers also wintered there, and Belgau said his older brother had gone to school with one of the Wallendas.

After a few hours talking to Belgau in his Port Angeles home, I wasn't quite sure what to make of him. Sweet and avuncular, to be sure, passionate and committed to his ideas, yes. "If I wasn't completely sure, right to the core of my being, that I'm right," he told me at one point, "I couldn't go out and have kids do this, no more than I could pick your pocket. When I'm dead and gone, I think I'll have made life better for a whole lot of kids." But just as with the psychiatrist Ken Erickson, who was long on anecdotal evidence but short on published studies, anyone with a shred of healthy skepticism would have to remain dubious about Belgau's system. And I was.

The further I explored the cognitive connection to balance, however, the more I uncovered independent strands and shards and snippets of evidence that all seemed to point to the possibility that Belgau, and others sailing the same waters, could be on the right tack. Was I finding these links because I wanted to believe the vestibular system has functions other than those traditionally ascribed to it? Or was I simply attracted to the idea that the brain is far more complicated and mysterious than we can conceive? That instead of becoming known, it seems to grow more complex the more secrets we learn, so complex—and yet so simple in a way—that people who aren't brain scientists, through sheer intuition and imagination, can come up with plausible ideas about how it operates?

Belgau was an educator and psychologist, grounded in the operating principles of aircraft engines, pendulums, and children who have trouble reading. Two other unlikely contributors (or psuedocontributors, depending on your point of view) to brain science, who would fall into the same camp as Belgau, and whose theories seem to bolster his, are a pair of retired computer scientists who made a second career in neuroscience research. Henrietta and Alan Leiner, who are married, in their early nineties, and living in California, are still pursuing answers to the many unconventional questions they began asking decades ago, which ignited a revolution in the way we view a part of the brain called the cerebellum.

Its name is Latin for "little brain." About the size of a fist, shaped like a bean, the cerebellum is located at the back of the skull, just above the brain stem. Though by weight it accounts for just 10 percent of the brain's mass, it contains more neurons than the rest of the brain and spinal cord combined. (The brain's total number of neurons is estimated to be around 100 billion.) The basics of cerebellar anatomy were outlined by Ramón y Cajal of Spain in the 1880s. Later, an Italian physiologist, Luigi Luciani, discovered that animals whose cerebellums were removed had trouble with equilibrium and coordination. Thus, since the latter part of the nineteenth century, scientists have believed that the primary role of the cerebellum is to control posture, coordination, and equilibrium.[2]

Dr. John Ratey describes how the cerebellum works:

Information about body movement and position enters the cerebellum, where it is processed. Instructions are then sent out to modify posture and coordinate muscle movement. This is more crucial than it may sound. For movements to be performed, the brain must know the position and speed of your body and of each limb and where you are in space and time. Spatial orientation and posture are essential to knowing 'where you stand.' The only reason you remain upright and don't fall down because of gravity is constant monitoring by the cerebellum. It adjusts postural responses at the brain stem, which sends messages down the spinal cord that control muscles that straighten and extend the torso and limbs, fighting against the downward force. This incredible feat is being accomplished all the time, without our being aware of it.[3]

All vestibular nerve signals travel initially from the eighth cranial nerve to one of two destinations: the brain stem or the cerebellum. Those that reach the cerebellum connect to the oldest part of that structure, in evolutionary terms, called the archicerebellum. The next oldest part of the cerebellum is the paleocerebellum, which governs proprioceptive inputs from the legs that help maintain an upright posture. The neocerebellum, the newest part to evolve, receives most of its input from the cerebral motor cortex and is associated with fine motor control, especially of the fingers.

Another role of the cerebellum, discovered in the 1970s, is in encoding the learning and memory of motor skills, especially complicated ones like riding a bicycle, walking a tightrope, or juggling. Without the cerebellum it wouldn't be possible, say, for someone who hadn't ridden a bicycle in forty years to climb aboard and roll away, as many people can. The memory of how to balance on a moving two-wheeled vehicle is locked into the cerebellum, according to this theory.[4]

The third and most controversial function of the cerebellum may be as a support system for certain cognitive tasks. Although the hard physical evidence has been made possible in the last few years by advances in technology—specifically functional magnetic resonance imaging (fMRI) and positron emission tomography (PET)—the theory was first conceived by the Leiners back in the early 1980s with only paper and pen and insight worthy of Edison.

For years their views were scorned and ridiculed by most of the scientific community. This chilly reception was likely the result of two factors. One was the deep entrenchment of the idea that the cerebellum was only a controller of motor activity, which had been the prevailing view for 150 years. The other was that the Leiners did not possess an acceptable pedigree. Both had started their professional careers as mathematicians, switching to the research and development of computer systems at the National Bureau of Standards. Alan went on to work at IBM for two decades, while his wife took time to raise their family. But when the children were grown, Henrietta, who had been intrigued by the similarity between the brain and electronic computers, decided to go back to school to study neuroanatomy. As a nondegree student at Columbia University, she dissected a human brain and was struck by the number of nerve fibers descending from the cortex into the cerebellum. Drawing on her knowledge of circuit design and information flow, she realized that the size of this bundle of nerves was significant. "When you dissect a human brain," she told the *New York Times* in 1994, "you can see with the naked eye an enormous cable of 40 million nerve fibers descending from the cortex to the cerebellum. I knew how much information it could send. I was just floored. I thought, 'This is some terrific computer down here.'"[5] To put Henrietta's statement in perspective, about 1 million nerve fibers travel from the eyes to the brain along what's called the optic tract.

Henrietta puzzled over this finding, and over the fact that primates don't have a proportionally similar-sized nerve bundle. If the cerebellum was useful only for motor control, as her professors told her, then why wouldn't primates, who have greater motor demands than humans by virtue of their gymnastic, arboreal lifestyle, have a similar "wiring" system? Looking further, she found that the most recently evolved section of the cerebellum, called the neocerebellum, is much larger in humans than it is in primates. Perhaps, the Leiners reasoned, the cerebellum — which fossil skulls indicated had expanded at least three times in the last million years, keeping pace with the growth of the cerebral cortex—had something to do with traits that were uniquely human. Language, logic, musical ability, planning—could the cerebellum be involved in these decidedly nonmotor abilities? Her professors shook their heads.

The first evidence that the Leiners might be right came from a neurologist named Robert Dow, at the Legacy Good Samaritan Hospital and Medical Center in Portland, Oregon. The Leiners had contacted him about their hypothesis because he was an authority on the cerebellum, having written a textbook on the subject. Dow was entranced by the Leiners' original thinking. When he located a patient with a faulty neocerebellum, Dow ran some tests. Instead of displaying motor problems, the patient had trouble with a cognitive challenge that required advance planning. It was a revelation to Dow, and in 1986 he and the Leiners cowrote a paper published in the journal *Behavioral Neuroscience* called "Does the Cerebellum Contribute to Mental Skills?"

"It caused a sensation," Henrietta told the *New York Times*. "People who had vested interests in the cerebellum as a pure motor device did not want to be challenged."[6] One of the people who rolled their eyes when they read it was Peter Strick, now at the University of Pittsburgh, where he is a professor of neurobiology, neurosurgery, and psychiatry. "I thought they were nuts," Strick told the university journal *Pitt Med* in 2002.[7]

But in the early 1990s, Strick's own research turned him into a believer. He used two different types of experiments to corroborate the Leiners' views. One involved a novel method of tracing connections between neurons in primate brains using mild viruses. The virus moves between neurons, mapping regions of the brain where signals travel. When Strick inserted a virus into a monkey's cerebellum, it traced a path to a part of the prefrontal cortex used in memory and higher executive functions. Further testing showed that this connection was a two-way loop, allowing information to continuously travel between the two structures. "The most important thing [Strick] has contributed," Henrietta told *Pitt Med,* "has been to show for the first time that the cerebellum not only sends fibers and information to some prefrontal areas of the cerebral cortex, but also receives fibers from these same areas. No one had ever seen that before. That means the cerebral cortex can talk to the cerebellum and exchange information. That kind of feedback loop is central to the computer."

Strick and his colleagues used fMRI technology to monitor activities that caused the cerebellum to activate. When experimental subjects were asked simply to move pegs on a board with their hands, a part of the neocerebellum called the dentate nucleus was triggered, but only slightly. Then the subjects were asked to solve a difficult puzzle. As they sat thinking about what to do, bodies and hands motionless, their dentate nuclei "lit up like a Christmas tree," according to Henrietta Leiner, displaying activity three to four times greater than in the easier task. While Strick said he believed this study was proof that the cerebellum plays a role in cognitive tasks, he isn't sure exactly what that role is. Other scientists remain skeptical, believing that the cerebellum, despite the new evidence, is still just a movement coordinator. Perhaps, these scientists argue, cognitive functions such as planning ahead and sequencing are somehow associated with physical movement, and that's why the cerebellum is active during these forms of cognition. But the Leiners firmly believe that these studies confirm that the cerebellum helps the brain not only with physical agility but with mental agility as well. The cerebellum, they observed, "makes predictions (based on prior experience or learning) about the internal conditions that are needed to perform a sequence of tasks in other regions of the brain, and it sets up such internal conditions in those regions automatically, thus preparing those regions for the optimal performance of those tasks. By doing this, the powerful and versatile computing capabilities of the cerebellum would be used for providing automatic help to various other regions of the brain, helping them to do their work better."[8]

This theory sounded remarkably similar to Belgau's views about the vestibular system as an underlying support system for other senses and for cognitive tasks. It also reminded me of the puzzling discovery Jeffrey Taube made in his research on animal navigation. In route-finding situations in which the vestibular system didn't seem to be directly involved, it turned out that it was. When the vestibular apparatus was "turned off," the whole system mysteriously shut down.

Joining Belgau in his assertions about the importance of the vestibular system in cognition is a widely published New York psychiatrist named Harold Levinson. Since at least 1980, he has argued that reading and learning disabilities are caused by problems with the vestibular system and cerebellum. But instead of employing Belgau's prescription of physical exercises, Levinson says he has achieved similar results with drugs that combat motion sickness. (Besides advocating a pharmaceutical regime, Levinson also diverges from Belgau by claiming that 80 percent of common phobias, such as fear of heights and open spaces, are caused by vestibular disorders, and he claims success treating these maladies with a cocktail of drugs that includes anti–motion sickness medication.) Like Belgau, he maintains that the results he sees with his treatment methods are enough to convince him they're worthy, despite a lack of rigorous testing.

Levinson and Belgau have another thing in common. Both were involved, Levinson directly and Belgau indirectly,

in the astonishing success of a British company called Dyslexia, Dyspraxia and Attention Disorder Treament Centre, or
DDAT. Not only did the company adopt the pair's theories
and commercialize them, but it also had the theories scientifically validated.

DDAT's founder, Wynford Dore, is a fifty-eight-year-old
Welsh multimillionaire who made his fortune selling fireproof
paint. In 1994, his daughter tried to commit suicide, reportedly because of severe dyslexia that had plagued her throughout her life. Dore began researching methods that might help
her. On a trip to Hong Kong, he serendipitously picked up
one of Levinson's books on dyslexia at an airport and read
it on the flight back to London. He contacted Levinson and
flew him to London to conduct studies on its efficacy. Dore
was put off by Levinson's drug-based approach, so he turned
to researchers who devised an exercise-based program that
addressed the same issues. Based on and nearly identical to
Belgau's, the regimen involved tossing and catching beanbags
while visually tracking them, standing on one foot or on a
balance board, and other "stretching and coordinating exercises." After serving as a guinea pig, Dore's daughter was allegedly cured of her dyslexia. Dore realized he had stumbled
upon another "product" he could sell, one that might help
millions of people with learning disabilities, and he began in-
vesting some of his fortune to set up private DDAT clinics
throughout Great Britian (there are now seventeen, including
several in the United States and Australia).

I watched a dramatic demonstration of Dore's techniques on
a segment of the CBS television news program *60 Minutes II*.
It profiled a dyslexic thirty-year-old man named Martin Long
who had struggled since childhood with his problem and was
often called "stupid" by schoolteachers. CBS selected him at
random from among a group of people just starting DDAT

treatment. When a CBS reporter asked him to read a passage from a book, his attempt was tortured, full of hesitations and mispronunciations, much like the child on tape that Belgau had me listen to. Nine months later, after daily treatment, Long was again asked to read the same paragraph. This time the words flowed with almost impeccable smoothness. The day after the broadcast, however, the International Dyslexia Association (IDA), headquartered in Baltimore, issued a press release refuting DDAT's methods as "predicated on research that has been questioned by many neuroscientists...[and] not supported by current knowledge."[9] As to the alleged gains made by Long and others, the IDA stated that it "may be an extreme leap of logic to assume the [DDAT] treatment was solely responsible."

For the first few years after DDAT opened, Dore's claims of his program's success provoked scorn like the IDA's from most of the scientific establishment. First of all, as is typical with most ventures of this sort, it seems, no studies had yet been formally conducted to prove its worth. Critics also claimed that it wasn't clear exactly who would be helped and how long the effects, if any, would last. And there was the matter of cost: $3,000 per person covered a few beanbags and a balance board? Actually there was more to it. A detailed analysis of a subject's vestibular system was performed on the first visit, using an expensive, sophisticated machine you may remember from Karen Perz's vestibular physical therapy office. Called the Balance Master, it is made by a U.S. company called NeuroCom and was originally designed to analyze NASA astronauts. It provides precise data about the condition of one's overall balance and analyzes the contributions from the three sensory inputs: visual, proprioceptive, and vestibular.

In 2003, much of the air was expelled from the arguments of Dore's detractors when a peer-reviewed study provided the first solid evidence that his methods actually worked—and worked spectacularly well. It was titled "Evaluation of an exercise-based treatment for children with reading difficulties" and was published in the journal *Dyslexia*. Some critics downplayed the study for its small sample size and because two of the investigators, though well respected in their fields, were affiliated with DDAT. The results were nonetheless eye-catching.

Thirty-six students with severe dyslexia, from a primary school in Warwickshire, England, were selected for the study. Over the next year, the children were given six months' worth of twice-daily, ten-minute, parent-administered sessions of exercises per day. Neither group received any other remedial assistance during the study. Their results, when compared to a control group and to their own performance on literacy tests the previous year, were remarkable. The improvement in their mean "reading age," a marker of reading proficiency, was three times greater than it had been the previous year; reading comprehension was five times greater; and progress in writing was seventeen times greater. Compared to the control group, they had "significant improvements" in cognitive skills, including phonological skill, reading fluency, dexterity, and verbal fluency. There were other benefits as well, though it was no surprise that most of the kids improved their balance significantly, as measured on the Balance Master.

The investigators concluded their report by stressing that the study was small and in the future needed to expand its sample size "to confirm these preliminary findings and to explore the ways by which the exercise mediates the literacy improvements. Nonetheless, the results do suggest that the exercise treatment was effective, not only in its immediate target of improving cerebellar function but also in the more controversial role of improving cognitive skills and literacy performance."

Were these just temporary gains? Follow-up testing showed that they were not. A year after the program, the subjects' improved abilities were still evident, indicating that the effects could be permanent. Even more interesting was that those who seemed to benefit most from the program were students with the most severe dyslexia. A year later, their improvement was reportedly so good that they would no longer even qualify to take part in the study. And what about attention disorders, which after all were part of the company name? How effective was the program in handling this supposedly related problem? Standard testing (though outside the scope of the study) revealed that 50 percent of the subjects in the study "had sufficient criteria to have an attention deficit/hyperactivity disorder diagnosis," according to the DDAT medical and research director, Dr. Roy Rutherford. A year later, an astonishing 0 percent fit this category.

And on and on. The results sound almost too good to believe. But no one has come out and said the study was rigged or falsified. No scandals have hit the tabloids yet, though the British press has lavished attention on Dore since the study was published, bringing even more attention to his company—and more customers through his doors (between 2002 and 2005, about 20,000). While it's perhaps too early to make any grand proclamations from such a small study, it does seem to provide evidence that there's an association between the cerebellum and cognitive functions. As the study authors noted, "If the Cerebellar Treatment Hypothesis does indeed turn out to be even partly valid, then this could revolutionise the treatment of literacy in dyslexia. Rather than attempting laboriously to scaffold the inefficient learning processes of the dyslexic child, one would first treat the cause (the cerebellum) and then learning would take place relatively normally—in all areas of skill."[10]

Belgau, left out in the financial cold in all these proceedings—he hasn't earned a dime from Dore because his methods or products weren't copyrighted or patented—does feel somewhat vindicated by the *Dyslexia* study. It's the sort of study he would like to have done years ago but could never afford. "It's nice to have the validation and those things," Belgau told me. "It's good to have that. I am seventy-two and I've lived a good bit of my life where people looked at me and said I was kinda funny." He hadn't yet had a chance to read the study, but he'd already formed an opinion of one of its tenets, that the cerebellum is where the retraining takes place. "That's too narrow," he says. "When you sell something, you have to narrow it down. It's a whole lot more than that. The cerebellum is just part of it. Primitive brain structures: the spinal cord, brain stem, limbic system, and the cortex. All of these pieces. What happens is they all work together. You really can't separate them."

Belgau's view was echoed by one of Peter Strick's research investigators at the University of Pittsburgh, Richard Dum, an associate professor in the neurobiology department: "The conventional view of the brain used to be one of localization," he told *Pitt Med*. "This part of the brain controls movement, this part controls thought, this part does feelings. They each have their own discrete roles. But there is a lot more interaction and cooperation going on than was once believed. We have seen that the brain is much more complex than people thought."[11]

In the end, Belgau wasn't completely forgotten. An American entrepreneur who helped Dore set up DDAT noticed that Belgau's methods had been "borrowed." He contacted Belgau, and the two agreed to form a partnership. The entrepreneur would help Belgau market his program more effectively in exchange for a percentage of the profits. Belgau, at least on the surface, doesn't appear to harbor grand ambitions for his company's future. Rather than modernizing his instructional videos or contracting with a public relations company, he is more interested in things like going down to the shop every day and getting sawdust on his pants as he cuts yet another balance board out of blemish-free Russian plywood. But if Belgau seems content to let others carry the torch now, the same can't be said for a group of scientists employing advanced technology to improve the lot of people with severe equilibrium problems. With inventions straight out of the pages of science fiction, the scientists we'll meet in the next chapter are bullish on the prospect of allowing the Cheryl Schiltzes of the world to lead more balanced lives.

Chapter Ten

New Balance

Miracle isn't a word scientists are fond of using. First of all, it doesn't sound very scientific. Its hint of voodoo and mysticism makes other scientists shudder. In a scientific experiment, the word is evoked only when something unexpected and unexplainable happens. That's precisely what prompted Yuri Danilov, a neuroscientist at the University of Wisconsin at Madison (UW), to describe as miraculous some of the effects of an experimental vestibular prosthesis on its first human subject, Cheryl Schiltz. For her part, Cheryl certainly wasn't expecting miracles when she was recruited for the experiment by the prosthetic device's inventor, a UW physician and professor of rehabilitative medicine named Paul Bach-y-Rita. She didn't have any expectations at all when she first began using the device in 2000, other than a slim hope that somehow her condition might improve. But in retrospect maybe it wasn't so surprising that the device could yield miraculous results, for in the forty or so years since Bach-y-Rita built the prototype he had used the strange-looking machine to prove the concept of what's termed sensory substitution, and in the process, albeit temporarily, it had returned sight to a blind man and touch to a leper.

Sensory substitution, the idea that one sense can stand in for another, is not new, according to Bach-y-Rita. When a blind person uses a cane, he's creating spatial knowledge of what's around him through tactile sensations received at the tip of the cane. The same thing happens when "reading" Braille, in which fingertips substitute for eyes. Ordinary reading, says Bach-y-Rita, was perhaps the first sensory substitution "system" because it is an "invention" devised by people "that visually presents auditory information," the spoken word. Back in the 1960s, Bach-y-Rita thought it might be possible to create a "human-machine interface" that enables sensory substitution. Using "junk pile" equipment because he had difficulty funding the experiment, he and four colleagues at Smith-Kettlewell Institute of Visual Sciences in San Francisco outfitted the back of a discarded dental chair with a grid of four hundred tiny electrical stimulators. A television camera sent images to an electronic device that translated the input into weak electrical pulses. Six blind subjects, who sat in the chair with their bare backs against the stimulators, were taught how to interpret the signal patterns as they maneuvered the camera. They learned how to discriminate between different types of lines, combinations of lines, and solid geometric forms. Then they progressed to recognizing twenty-five common objects, such as a toy horse and a telephone. After several hours of training, the subjects were able to "discriminate between individuals, to decide where they are in the room, to describe their posture, movements, and individual characteristics such as height, hair length, presence or absence of glasses, and so on," according to a paper written by the scientists and published in *Nature* in 1969.[1] Without prompting by the researchers, the test subjects reported that the sensory information seemed to come from the front of the camera, rather than from the stimulus on their backs.

But Bach-y-Rita's experiment was received mostly with disdain by the scientific community. Part of the reason for this response may have been that much of the data was anecdotal, not the hard quantifiable information science demands. According to Bach-y-Rita, the Zeitgeist wasn't in place to receive such a radical idea. "There were a lot of studies showing how incredibly complex the visual system was," he said, "and yet you come along with a four-hundred-point array stuck someplace on the body and people can see with it. It just was too far out, just too far out." Another contributing factor was that the equipment at that time was too heavy and cumbersome to be practical, so there didn't seem to be a legitimate use for the technology.

Bach-y-Rita was undaunted. For him the experiment was an elegant demonstration of a neuroscience principle that had intrigued him for years: the brain's inherent "plasticity," the extraordinary capacity to change its structural organization and function when necessary. In the case of blind people, they have lost the ability to process light through their eyes but haven't necessarily lost the ability to see, "because we do not see with the eyes," he said, "but with the brain." In people with normal vision, the optical image doesn't go past the retina but is transformed into nerve signals and directed to the brain for processing. His experiment proved that the nervous system could adapt to a different form of visual sensory input (via the television camera and tactile stimulators), interpreting it with a high degree of accuracy. PET scanning would later show that the visual cortex was being activated by these tactile signals.

The same sort of result was achieved with a man afflicted by leprosy who had lost the tactile sense in his hands. Bach-y-Rita and his colleagues made a glove for the man to wear, with a pressure-sensing device on each finger. The electrical signals from the glove were routed to a computer and then transmitted to a small tactile stimulator on the man's forehead. "After an hour of introducing him to the system," Bach-y-Rita said, "I sent the guy home and said, 'Come back tomorrow and tell me what you can do with it.' And he came back and explained that he could perceive the cracks in the table and the texture, and then he turned to me and said, 'You know, doctor, the most extraordinary thing was touching my wife's face, it was the first time in twenty years I could feel it when I touched her.' So here within a couple hours of use he was already spatially locating, and he was not perceiving it on the forehead, he was perceiving it where his brain sent his hand to pick up the information."

Despite these early spectacular successes, it would take another thirty-five years for the technology, and Zeitgeist, to evolve far enough to allow Bach-y-Rita's "human-machine interface" to win wider acceptance. Advances in instrumentation and computer electronics finally made it possible, in the 1990s, to build a user-friendly sensory substitution machine—now called the BrainPort as it approaches its commercial debut—far more compact than the original. One of the biggest changes from the original was the site of the stimulus. After experimenting with almost every part of the human body, Bach-y-Rita settled on the tongue as the ideal location, mainly because saliva conducts electricity easily, dramatically lowering the voltage required compared to dry skin. (This site may sound painful but isn't; the sensation resembles the fizzing of champagne bubbles, according to one experienced user.) The first application of the new version was again for the blind. It consisted of a head-mounted digital movie camera connected to a computer, and a flat cable that traveled up and into a user's mouth, where it ended in a miniature array of 144 stimulus points. The machine worked so beautifully that one subject was able to see a ball rolling toward him on a table, reach out with his hand, and catch it. He could also navigate down a hallway without any additional assistance.

But again, it was hard to quantify the results. So when Mitch Tyler, one of the principle researchers in Bach-y-Rita's lab, suggested modifying the device so that it could augment a faulty vestibular system (after suffering a bout of dizziness himself due to an inner-ear infection), his colleagues seized on the idea. Here was an opportunity to measure the device's effects more easily because it would be a simple matter to test a subject's balance abilities before, during, and after sessions with the BrainPort. Instead of a head-mounted movie camera, this version would use a helmet filled with special motion-sensing equipment. Inside an ordinary plastic construction worker's helmet, the scientists attached small accelerometers, sensitive instruments that measure acceleration forces caused by movement or gravity. (A common use for accelerometers is in automobiles. Upon detecting rapid deceleration of a vehicle, as in a crash, the instrument signals the air bags to deploy.) In the BrainPort application, accelerometers measure tilt as a carpenter's level does. And similar to the way the vestibular system's own otolith organs work, it can also gauge linear acceleration. When a person wears the helmet, every nuance of both linear motion and tilt is converted to electrical signals and transmitted to a computer. Here the signals are translated into an electrical pattern and relayed through a cable to a postage stamp–sized grid of stimulators that sits on the tongue. If the subject tilts to the left, for instance, stimulus is felt on the left side of the tongue. When a subject's head is still, the stimulus also remains stationary, indicating total balance. After a period in which the subject learned to "read" the signals, it was hoped that the BrainPort would help a person maintain an upright posture and also reduce the probability of falling.

As the team prepared the prototype for testing, they began looking for a test subject, ideally someone with severely compromised vestibular function on both sides. Bach-y-Rita contacted an otolaryngologist at the university and asked him for appropriate candidates. Cheryl Schiltz was among them. She was ideal because not only was she local, but her uncommon disability, bilateral vestibular disorder, was particularly debilitating. And though the extent of the damage to her vestibular system wasn't known at the time, it was obvious that the effects on her behavior and function were severe. "She came leaning against the wall, with two canes," Bach-y-Rita says of their initial meeting. She was so wobbly in fact that even the act of sitting was precarious; unless a chair had arms she was liable to fall out of it. During the interview, she outlined the enormity of her suffering, all the cognitive and behavioral symptoms—sleeplessness, memory problems, and the inability to multitask. Her obvious intelligence and articulate nature were also appealing to the researchers, so she was an easy choice for the team.

The first part of the BrainPort trials, which began in 2000, were all done with Cheryl in a seated position. "What we noticed," she said, "was that with my eyes closed, without the system in my mouth, I had great difficulty keeping my body still. [But] when we put the tongue display in, and I had that sensation to maintain and to concentrate on, I didn't move!"

The team extended the time she spent on the BrainPort from one hundred to two hundred to three hundred seconds, and they noticed a curious phenomenon. When she took the tongue display out of her mouth, she was able to remain still for a few moments, as though the effects lingered in her brain. After a couple of months of sitting trials, Cheryl began lobbying for faster progress. She suggested they move to standing trials. This was the big moment, to see if the BrainPort would stabilize her wobbly stance, something she'd been unable to do since her vestibular system was destroyed in 1997. Her standing balance without the device was "all over the place," she reported, but seconds after she put the tongue display in her mouth she could feel her gyrations grow smaller. Oh my God!, she thought to herself. Even with her eyes closed she could maintain an unwavering position. Overcome with emotion, she felt tears well up in her eyes. She had her balance back, even if for only a few minutes.

"By clinical analysis she had one hundred percent functional loss," says Yuri Danilov, the lab's chief scientist and director of clinical research. "By definition she had no hope, no future, because nobody knows how to recover damaged sensory cells. Okay, we put her on artificial feedback, we provide her with an artificial signal from an artificial sensor, and as expected we see [her] use the signal to stabilize [her] body. That's no miracle." But what came next surprised everyone. Someone suggested increasing the duration of a session to twenty minutes. Afterward, as Cheryl took out the tongue display and removed the helmet, she felt strange. Something had changed. Looking up at the ceiling, which usually would have made her feel dizzy, evoked no response. "The first thing I noticed was the lack of noise," she said. "What I mean by noise...it's not an audible noise, but it's like...an internal electrical kind of noise. It was totally quiet. I just felt relieved. I could feel

relaxed and comfortable.... And I turned and I looked at
Mitch [Tyler], and I said, 'Something is different.' And he said,
'What do you mean?' and I said, 'I feel fine. I do not feel out
of balance.' I didn't have any shoes on, and I just took off
and started running through the building. And I ran out in the
parking lot. I felt like I was touched by the hand of God and
cured. I felt normal, whatever that definition is. It was unbe-
lievable. For a full hour I was like that. And for me that was
the most incredible thing ever...for an hour I felt normal! I
could close my eyes. I could spin around in circles, I could do
whatever I wanted to." The residual effects lasted until that
evening. "I got that heavy feeling again, my eyes got kind of
dry, and the fatigue set in again. And the next morning, I got
up and it was horrible. It was just like day one of being damaged. I felt like something was worse, and I called Mitch, and
I said, 'I don't know what's wrong, but I feel worse than ever.
I can't see anything, I'm so fatigued, this is so bad,' and I was
crying, and I said, 'Did we do something bad?'"

But further twenty-minute rounds on the machine increased the amount of time she experienced "normal" vestibular responses. Little by little, she built up to a half day,
then a full day. Eventually two twenty-minute sessions on the
BrainPort a day, one in the morning and one in the evening,
allowed her to remain virtually symptom-free. At one point
many months later, she was able to go four months after her
last BrainPort treatment, during which some of the symptoms returned, sleeplessness and fatigue, but never close to
the state she was in before she began using the machine. This
kind of effect had never happened before in tests of the Brain-
Port, with either blind subjects or people with tactile loss.
The question of whether Cheryl alone was able to experience
residual effects was answered later, after trials using twenty-
four other people with vestibular deficits produced the same

result. "Our patients not only recover ability to keep posture, keep verticality," Danilov said, "but they recover ability to do whatever they want. They can ride a bicycle, go walk a beam, jump rope, dance, stand on one leg"—all *after* removing the BrainPort. This was the miracle.

"The results turned out to be more extraordinary than we expected," Bach-y-Rita explained, "because we were originally working on just straight substitution. It works when you use it, and you take it out and it doesn't work. And yet we've found that using it for twenty minutes could give up to three, four, five hours of complete normality. We didn't predict this would happen. People riding a bicycle, climbing stairs, and whatnot. That's been hard to explain physiologically."

Other scientists, hearing about the therapeutic effects of the machine, were naturally skeptical. Dan Merfeld, the Harvard vestibular research scientist, was among them. "Yeah, that has surprised a lot of us," he said. "Obviously the next step is to try to find out what is the cause. To my knowledge, nobody has really come up with a successful explanation. And I think because we're not at that stage, it makes us all wonder: is it real or is it some artifact? Placebo effects have been observed in many strange ways in science throughout the years. I'm not even suggesting that this is a placebo effect, but because placebo effects have been observed, it forces us to at least consider that until we can explain it. All of the lingering effects are somewhat of a surprise and just show us, if they hold up, how much we have to learn."

Another incredulous onlooker was Owen Black, the venerable vestibular scientist from Portland, Oregon. He was intrigued enough by the UW lab's findings that he invited Tyler and Danilov to bring the BrainPort with them to Portland for a week of further testing. He threw some really hard curves at them, according to Bach-y-Rita, hooking up the machine to patients with intractable cases of Ménière's disease, which is different from the bilateral vestibular disorder that Cheryl Schiltz had, with different symptoms. Two of them had not been helped by traditional therapies but appeared to respond favorably to the BrainPort almost immediately. Black was impressed and decided to continue working with the UW group. "It's a phenomenal piece of equipment," Black later told me, "but we don't know how it works."

Since those first heady days, several hypotheses have surfaced to explain the "miracle." Bach-y-Rita published a paper twenty years ago in which he talked about the brain's amazing ability to reorganize itself with as little as 2 percent of the surviving tissue in a particular system. "When a person loses the vestibular system, from gentamicin, let's say, there's never a hundred percent loss," he explained. "Let's say it's a fifty percent loss or a ninety percent loss, but that remaining tissue can be reorganized, totally, to take over the functions of whatever has been lost. And that's maybe the most exciting finding of this whole thing." Though it has a great deal of merit, Danilov says this idea is hard to investigate because the only way to determine how much tissue remains requires a sampling procedure that could be life-threatening. Nevertheless, he says, "there is still high probability that you have residual tissue." This residual tissue would be critical for regaining balance with the help of the BrainPort but would not be sufficient on its own because its very weak signal would disappear among the brain's general noise (electrical activity between and along neurons). "What our device is doing, and we have pretty good evidence, is suppressing the noise in the brain, and actually improving signal-to-noise ratio," Danilov explains. "As soon as the brain catches this weak signal, it starts to recalibrate all circuitry in the brain. That includes motor control, visual control, everything that's involved in the vestibular system....So from that point of view it's not a miracle. It's not a pill that we're giving, it's not a screwdriver to tune up something. No, we're just giving the brain the opportunity to use internal resources."

Exactly how this occurs, though, will probably take years of research to explain. Nevertheless, seeing is believing, and Cheryl's early dramatic success had a profound effect on the BrainPort project. Suddenly a floodgate of funding opened up from various sources, though Bach-y-Rita mentioned one National Institutes of Health reviewer who turned down a grant request recently because he flat out didn't believe the reported achievements of their initial experiments. The Discovery Channel, the *Today* show, and *National Geographic* came calling to Madison after the *New York Times* ran a feature on the BrainPort in 2004.[2] The momentum helped Bach-y-Rita create a company to build a commercial product, which is now (2007) undergoing Food and Drug Administration trials in the United States (though it was approved for use in Europe and Canada in 2006).

A number of different applications for the BrainPort are in development. One hallmark of the machine's underlying technology is its almost limitless adaptability. "The device itself is a universal interface," Danilov explained, "basically a universal sixth sense." Besides the application for vestibular and vision substitution, fifty-four other potential uses have been outlined. One is to help deaf people improve their ability to read lips; another is to aid navy divers navigating in murky waters, perhaps by converting information from sonar sensors into electrotactile stimulus. "You can use the Brain-Port to replace any sensory system," Danilov says.

The BrainPort facilitated the recovery not only of Cheryl's damaged vestibular system but of her life. Bach-y-Rita, during the experiments, became like a father to her and convinced her to study and major in rehabilitative medicine, which she embraced as though she had been born to it. Having lived with a disability herself, she feels she is uniquely qualified to help others deal with physical or mental deficits. Bach-y-Rita also offered her a job in 2003, writing her into one of the BrainPort grants, and she worked as the clinical coordinator and office manager. Because she had been unable to find work after her vestibular injury, she had been forced to rely on Social Security disability checks until Bach-y-Rita's offer. "One of my happiest days," she said, "was when I called the SSDI [Social Security Disability Insurance] people and the guy answers the phone and says, 'Well, what can I do for you?' And I say, 'I don't want your money anymore.' And he goes, 'Excuse me? I've never had anybody tell me that.'"

Since the BrainPort sessions began, in addition to regaining stable upright posture and vision, Cheryl also believes her cognitive impairments have receded, although there are no hard data to back up this claim. "My mood has increased, my sleep has definitely gotten better," she said. "I'm much more observant. I'm clearer, my fine motor skills are better. I can multitask." This anecdotal evidence that the BrainPort can influence psychology as well as physiology has many scientists puzzled. But Danilov, for one, sees this as an exciting research opportunity. "One of the problems with BVD [bilateral vestibular disorder] patients is cognitive loss," he explains, "an inability to work in a multitasking environment, inability to focus attention long enough, to understand text or the tasks that they're doing. It's a big deficit. It's something that nobody knows how to treat. Yet all our patients self-report a huge cognitive improvement."

Some scientists believe these sorts of cognitive impairments are due to the brain having to devote "cortical" or upper-brain resources to a "subcortical" task, such as maintaining posture.

This siphoning of resources may be diminishing cognitive performance. But others, as I outlined in chapter 9, think there may be a more intimate, though still poorly understood, connection between the vestibular system itself and cognition; with vestibular dysfunction comes cognitive deficiencies. The experiments with the BrainPort may provide an avenue for investigating this link. Danilov offered two theoretical guesses about how it might work. In what he called the "bottom-up approach," the vestibular system may be more deeply involved in other neural processes than just balance and visual stabilization. By repairing the vestibular system, these other functions are also improved. The second approach he described as "top-down." The BrainPort may somehow be "kicking, or waking up," a slew of homeostatic processes within the brain, in essence helping a faulty system regain its own internal equilibrium. In this view the vestibular system itself is somehow returned to a state of balance, allowing it to function properly.

Around the world, there are at least two other vestibular prosthesis projects under way, though the BrainPort is probably the closest to going public. One is a joint effort between the Oregon Health and Science University's Neurological Sciences Institute and Italy's University of Bologna. Its fundamental design is remarkably similar to the BrainPort, using accelerometers to measure body sway. But instead of transmitting the information to the tongue, it sends it in audio form to a set of headphones. "Different tones and intensities tell subjects when they are leaning outside of their central safe zone," said Marco Dozza, a graduate student in bioengineering at the University of Bologna, in a press release. "In addition, the sounds tell the subjects which way they are leaning so they can immediately correct the problem before they fall. For example, when subjects lean forward, they hear a high-pitched tone that becomes higher and louder the farther they lean forward. If subjects lean backward, they hear a low-pitched tone that drops and gets louder as they lean back. In addition, the tone becomes louder in the left ear when a subject leans too far to the left. The tone becomes louder in the right ear when the subject leans too far to the right." After the initial trial of the device in 2004, on nine subjects like Cheryl, with bilateral vestibular disorder caused by ototoxic medications, the subjects had more stable posture while wearing the device, but there was no reported evidence of a residual, or therapeutic, effect.

Another prosthesis is being tested by Conrad Wall at Harvard's Jenks Vestibular Physiology Laboratory. Now in an early stage of development, the device incorporates a head-mounted gyroscope as well as accelerometers, transmitting signals to vibrating stimuli placed on the abdomen. Wall stated that experiments so far have demonstrated positive results for controlling posture and sway, with further tests being conducted on effectiveness while a subject is walking. Wall believes his device may demonstrate a therapeutic effect, "but we are being conservative about what we claim, until we have a complete data set. In my opinion, device trials must control for the effects of learning a new task independent of any effects due just to the use of the aid. I have not seen peer-reviewed data [for the BrainPort] that account for these controls."

Wall and his colleague, Merfeld, are working on another interesting project at Harvard, a vestibular implant. Like cochlear implants for the profoundly deaf, a vestibular implant would be attached to the mastoid bone (directly behind the ear). It would consist of a tiny package of motion sensors, with an electrode traveling into the inner ear that would stimulate the eighth cranial nerve (which carries hearing and vestibular data from the inner ear to the central nervous system). The implant would in effect bypass the faulty vestibular signal.

The device is still only a concept, several years away from a prototype, but Wall reported that he has been collaborating with Swiss surgeons who have begun to prove the validity of the implant's design. Experimenting on human subjects, the surgeons have electrically stimulated the vestibular system in a way that provokes "robust reflexive eye movement in response," according to Wall. Merfeld explained that the ability to stimulate the vestibular ocular reflex would be one advantage of an implant over a sensory substitution device, as experiments with the latter have shown no effect in this arena. Presumably, patients with poor visual stability would be helped more with a vestibular implant than they would with a sensory substitution device. The big drawback, however, is that it would require surgery, whereas a prosthesis wouldn't.

Though it seems more far-fetched than an implant, another futuristic balance-enhancing device actually has solid research behind it, and perhaps a far greater number of people for whom it might be useful. At first blush, it resembles a gadget lifted from a rerun of *Get Smart,* the 1960s television spy comedy. Like Agent 86's shoe phone, this odd invention sits beneath the feet, embedded in an insole. It is the brainchild of James Collins, a professor in the department of biomedical engineering at Boston University and a recipient of a MacArthur "genius" award. Emitting a randomly generated "white noise" signal, at about 100 Hz, the device improves proprioception in the feet by 5 to 30 percent. Surprisingly, it appears to help younger people as well as seniors, according to a study published in the *Lancet* in 2003. Just like the BrainPort, it allows the brain to make better use of a weak signal.

Collins first came upon the idea in 1994, while attending a conference in Montreal. There he met Frank Moss, a physicist from the University of Missouri at St. Louis, who told him about a paper he had recently published about a phenomenon observed in the tails of crayfish called stochastic resonance (SR). *Stochastic* simply means "randomly generated," and *resonance* refers to "noise," which *Webster's* unabridged dictionary defines as an "electrical disturbance in a communications system that interferes with or prevents reception of a signal or of information, as the buzz on a telephone or snow on a television screen." The concept originated with physicists in the early 1980s who were trying to develop mathematical models that would account for the periodic occurrence of ice ages on Earth. They found that by introducing random fluctuations, or "noise," to their equations, the models would align with available geophysical data and explain the climatic changes. In the case of crayfish, Moss and his colleagues demonstrated that SR enhanced the detection abilities of sensory cells in the creature's tail. Instead of interfering with those signals, as one might expect, SR actually improved their function. This result was surprising to some people.

"It's a counterintuitive notion," Collins explained, "because in general we view noise as being a detriment or a nuisance for signal detection and information transmission. For example, we tune static out of our radios, we'll pay extra for cell phones that get good reception — noise-free or low-noise — and in engineering departments we teach students how to filter out noise to improve the performance of information-based systems. But what the noise effectively does is combine with the signal and enables it to get up over a threshold. At low levels of noise, there's not enough of the signal being boosted up over the threshold to be detected. And at high levels of noise, most if not all of the signal is being boosted up over the threshold, but the important features or information of the signal are swamped out by the very large levels of noise that have been added to the signal. And so there is some intermediate level that will suitably boost the signal up over the threshold without significantly contaminating the features of the signal. That really lies at the heart of the stochastic resonance phenomenon."

Moss challenged Collins to come up with a medical application for SR. On the plane ride back home to Boston, Collins mulled over the idea. He knew that proprioceptive neurons break down or diminish with age, disease, and injury, so he wondered if he could improve the function of these cells. He also knew that all neurons have a certain action threshold: below this level they don't fire; above it they do. He thought it might be possible to artificially introduce noise into the human body that would in effect lower the threshold of sensory neurons and thus boost their sensitivity.

When he got back to the lab, he began doing computer model studies to show that the general effect could work. He then moved to animal studies and eventually to trials using a specially designed insole. The *Lancet* study involved twenty-seven people, young (mean age: twenty-three) and old (mean age: seventy-three), all healthy and without any obvious balance difficulty. First their standing body sway was measured by a video camera, which tracked the movement of a reflective marker attached to the subjects' shoulders. Then the participants stood on vibrating gel insoles with their eyes closed. The SR signals were adjusted so that they couldn't be consciously felt; in other words, the subjects didn't know when the stimulus occurred. Sway was again measured and compared to the baseline data, revealing that it was significantly reduced when standing on the vibrating insoles. Although the younger subjects' sway parameters were lower when using the insoles, the greatest effect occurred in the elderly. In two aspects of sway (side-to-side movement being one), the elderly subjects' "performance" matched that of the control group of twenty-somethings.

A fountain of youth for balance? The science-fiction nature of the concept created a minor stir in the media and was mentioned prominently in the announcement of Collins's 2003 MacArthur award. He believes the insoles could at least reduce the chance of falling among the elderly and perhaps even enhance balance abilities in athletes of any age. In addition, another study, published in the *Annals of Neurology* in 2006,[3] showed the same standing-balance improvements in subjects with diabetes who had neuropathy of the feet, and in subjects with stroke, who often suffer from balance problems. Brain-injury and Parkinson's disease patients are two other groups who might benefit from the technology.

A company called Afferent Corporation has been working since 2000 to create a commercial product based on Collins's patented design. The company is still about a year away from having a product ready, reports Afferent's president, Jason Harry. Studies remain to be completed that the company hopes will show a benefit during everyday activities, like walking, climbing stairs, and getting in and out of chairs. Another hurdle is how best to package the components. Will they all fit inside a shoe or will part of the device have to reside in a separate box, perhaps strapped to the ankle? "The fundamental notion," Harry said, "of boosting sensory function in a beneficial fashion has absolutely been proven in human subjects. But it needs to pass muster as clinically relevant and something truly useful to patients and clinicians, and I'm quite confident that we'll get there."

If Collins's insole can improve balance by boosting proprioception, another angle of attack is to slow or reverse the natural decay of the vestibular system. As the body ages, hair cells within the semicircular canals and otolith organs, whose stimulation informs the brain about the head's orientation and movement, die off without replacing themselves. The more hair cells that are lost, the less sensitive the vestibular system becomes. The same goes for the auditory hair cells lining the organ of Corti in the cochlea. These hair cells also respond to movement, the movement of air, which triggers sound signals to the brain. Among the 32 million Americans[4] with significant loss of hearing, the most prevalent cause, by far, is hair cell loss. Age, ototoxic chemicals (such as gentamicin), loud noises, and certain diseases cause balance and hearing functions to deteriorate due to the permanent loss of these hair cells. So one of the grails of vestibular and hearing research right now is to find a way to spur the growth of hair cells to replace those that die.

One of the pioneers in this young and expanding field is Edwin Rubel, a professor at the University of Washington's Virginia Merrill Bloedel Hearing Research Center. In the mid-1980s his research group, along with another team at the University of Pennsylvania, made a startling discovery. While it was known that adult forms of lower vertebrates, sharks for instance, could regenerate inner-ear hair cells, the same wasn't true of higher vertebrates, birds and mammals. But in experiments on bird cochlea, both teams were amazed to find that, after damage to the bird's inner ears (with either noise or ototoxic drugs), new hair cells grew back. It was eventually learned that all vertebrates except mammals regenerate these cells. (The reason mammals have lost this ability isn't clear.) So the race was on to figure out the mechanism by which this regeneration occurred in birds, and, if possible, how to apply it to humans.

Recently, several groups of researchers have demonstrated that a small amount of cell division can be induced in the vestibular systems of mammals—rats, mice, and guinea pigs. This discovery was encouraging because it showed that the first stage of hair cell regeneration was possible. But because of a lack of funding, Rubel thinks it will be ten years before the concept might be proven in animals, and another ten years before hair cells could be regenerated in humans.[5]

A team led by Yehoash Raphael of the University of Michigan Medical School announced a big leap forward in this research in 2005. Working with guinea pigs, researchers were able to "turn on" a gene, normally active only in the animal's embryonic stage, that governs auditory hair cell generation. They had done this two years before, in what Raphael hailed as "the first time anyone has shown new hair cells can be grown in a mature mammalian ear."[6] But he hadn't been able to show that the new hair cells could rewire themselves to the brain or detect sounds. In the most recent experiment, scientists destroyed the animals' cochlear hair cells with ototoxic chemicals, causing deafness. Then a virus carrying the gene responsible for hair cell growth in mouse embryos, originally discovered in the fruit fly, was placed inside the left inner ears of some of the guinea pigs. Eight weeks later, microscopic investigation showed hair cell regrowth in the left inner ears of the treated animals, but not in the untreated ears. Further testing showed an ability to hear, albeit in a "distorted" fashion, in those "engineered" ears. Along the same lines, researchers at Harvard Medical School in 2005 reported success in neutralizing a gene in mice that causes ear hair cells to stop generating after birth. If this process could be replicated in humans, hair cells could keep reproducing and, theoretically, hearing could be maintained.

Short of a miraculous new discovery in hair cell research, Cheryl Schiltz will have to make do, at least for the foreseeable future, with the BrainPort to stay in balance. There are no permanent cures to serious balance disorders on the horizon. "You have to be realistic," Danilov told me. "It's impossible to grow up new tissue. Unfortunately, nobody has proved that you can regenerate sensory tissue. But right now, she's almost normal. She even refuses to consider herself a patient."

Cheryl is able to do things now that she could only dream of a few years before. "Hiking, going to Devil's Lake and playing around on these big boulders and rocks," she said. "Standing a foot from the edge. I thought my son was going to have a hernia: 'Mom, please don't do that.' I said, 'Chris, I'm going to do it. I'm not going to fall. I wouldn't do it if I thought I couldn't.' Believe me, I was scared shitless: Oh, this is really stupid, Cheryl, what if you decide to all of a sudden get that feeling?"

On the way back from the lake, she and her son came upon a creek with a log suspended above it, spanning its width. Cheryl told her son that she was going to walk across that log.

"No way, Mom, you shouldn't do that," Chris said.

His comment made Cheryl want to do it even more.

After stepping across stones in the water, which in itself was a feat, she clambered onto the log. A few feet across, she thought to herself, Okay, is this really one of the smartest things you've ever done?

And that was her mistake. She stopped and thought about it and froze. Chris came over and positioned himself beneath her. She reached out and put her hand on his shoulder. That's all she required, that tiny touch. She told Chris she needed to do this, and he nodded. As she started across the log again, Chris noticed that she didn't have a death grip on his shoulder. So when they were about five feet from the end, he stepped aside and said, "The rest is on your own."

She sucked in her breath and walked the rest of the way. Jumping off the log, she cried "like a baby." It was the first time she had attempted something so bold since her balance problems began.

"Every time I do something like that," she said, "a little more comes back to me, my life, and who I am, my adventurous spirit, all that is starting to come back alive. It's been swept under a rug for many years, and I'm not going to let it happen anymore."

It *was* a miracle.

Staying Glued to the Wire

One of the most arresting definitions of balance I came across as I was researching this book was one coined by a woman named Nancy Rowe, a Florida-based audiologist who uses vestibular therapy to help kids with learning disabilities. "Balance," she said, "is the action of not moving." I imagined a tightrope walker balanced on a wire. He appears motionless but really isn't. When standing on two legs, whether on a tightrope or on flat ground, our bodies are swaying almost imperceptibly, like the trunk of a Douglas fir in high winds, orbiting around our center of gravity. This subtle exterior action is mirrored within by countless electrical signals buzzing through the nervous system, transmitting information from the eyes, the vestibular system, and proprioceptive cells, deluging the brain with information about body position and orientation. The brain then automatically sends orders for muscles to fire in an all-out effort to keep us vertical. Thus the phrase *standing still* is something of an oxymoron.

It seems like a long time ago, that day my father was apparently standing still on a rock on top of Mount Si, lost his balance, and unwittingly propelled me into writing *Balance:*

In Search of the Lost Sense. A few months ago, we again hiked to the mountain's summit along with my brother and a friend. As I watched Dad maneuver among the rocks that day, my brother stood watchfully at his side, recognizing that his gait was a little tentative and wobbly. At one point, fearful of another fall, I made a sudden motion toward my father, as if I could somehow catch him if he wavered. I was glad when, after eating lunch and taking in the expansive view, we scrambled down off the peak and onto the smooth trail home.

Apart from the symbolism of my dad's fall—losing one's balance is like a little death, gravity catching you off-guard and pulling you down to the soil—I've thought a great deal about the physics and physiology of his tumble. Among the lessons it taught me was that, like balance itself, the cause of his fall was complicated. Part of the problem, as I've shown, had to do with the natural decline of his proprioceptive, visual, and vestibular functions. Then there was the novelty of walking among large boulders, an activity he does nowhere else. Consequently, his body hadn't had a chance to adapt to these very specific balance challenges. But the other element that seemed to be working here was something called height vertigo. It's a phenomenon many people experience when standing at the top of a building or cliff and looking out into empty space. You feel dizzy and a little disoriented, and your natural reaction is to move away from the edge to safer ground (a survival mechanism, perhaps). What's causing this uncomfortably unstable feeling is the absence of nearby visual reference points.[1] We're used to seeing markers—trees and fence posts and buildings and stair railings—as we navigate through the world, and they help stabilize us. When they're missing, the brain suddenly loses visual orientation, and the disconnect between visual signals and vestibular and proprioceptive cues causes us not only to get dizzy, but to increase our so-called sway parameters, the arc of our usually subtle swaying pattern. The farther we get from vertical, the less stable we become. So a double whammy hit my dad as he stood on that boulder years ago. His advanced age had increased the speed and arc of his sway, which alone would be destabilizing, but the effect was exacerbated by the lack of close visual cues as he looked out toward the valley below. It hadn't taken much—the simple act of removing his pack from his right shoulder—to pitch him off balance and off his feet.

Could my father have somehow avoided his moment of imbalance and remained standing on that boulder? In twenty years' time, will baby boomers like me possess better balance under similar circumstances? That depends on if and how we accept balance as a legitimate sense. When we're young, good balance is a given, and we don't need to pay it a thought. But as we move into our sixties, I think the evidence is clear that we can't afford *not* to think about it. Not just to prevent a potentially lethal fall, but to be able to continue moving gracefully through the world, to stay glued to the tightwire of life. We need to know what can disrupt our balance, how to adapt to failing faculties, and how to achieve the best balance possible. Like Robin Grindstaff, the ex-soldier with bilateral Ménière's disease, we have to devote part of our conscious attention to balance. Perhaps, as in Robin's case, that means moving more deliberately. Perhaps it means knowing when a hiking staff or cane would be beneficial. Certainly it means that we should attempt to slow the natural decline of balance. While there are many reasons to resist the relentless slide in our culture toward more sedentary lifestyles, maintaining a good sense of balance has got to be one of the least acknowledged yet most compelling.

If space research and vestibular physical therapy have taught us anything about balance, it's that the brain can adapt rapidly to altered states of balance, such as low gravity or diminished vestibular function. So it seems plain that, as we age, we need to allow the brain time to compensate for our naturally deteriorating balance faculties. The only way it can do that is to continually challenge the balance system.

It's not a difficult thing to do if you embrace the notion that balance is an important element of our well-being. It doesn't take a lot of money or time. The key, of course, is to stay active, and the more complicated your motion the better. Brisk walking, especially on uneven terrain, is ideal. Perhaps even more beneficial are dancing and aerobics classes, which demand even more balance.

Since I began writing this book, I've found several small ways to incorporate balance-boosting activities into my life. Taking a tip from one of the subjects of Dr. Steven Wolf's tai chi research, who reveled in his newfound ability to take off and put on his shoes while standing on one foot, I now find it useful, when I get out of the shower, to dry one leg at a time by lifting it up toward my chest while balancing on the other leg. Likewise, when I doff or don socks, I now usually stand rather than sit, balancing on one leg. I can feel the muscles and ligaments in my ankle firing wildly as they work to keep me upright.

What about the idea of challenging your balance as a way to keep mentally sharp? Though I've suggested throughout this book a connection, as yet unexplained, between the vestibular system and cognitive function, most experts make a point of saying that in a "normal" population (i.e., people with no obvious cognitive disabilities) good balance doesn't equal good mental performance. But that view may be about to change.

As I mentioned before, tai chi is among the most effective activities for enhancing balance. This choreographed system of movement is based on principles of Oriental medicine, which maintain that the mind and body are bound together inextricably, one always influencing the other. In traditional Western medicine, however, mind and body have usually been considered separate entities. A recent study[2] by Seattle researchers lends support to the Oriental view—and to the importance of maintaining good balance as we age. More than two thousand men and women over sixty-five, who had no signs of dementia when the study began, were evaluated every other year for six years on basic physical and mental abilities, including standing balance. At the end of the six-year period, the first signs of dementia in the three hundred or so who contracted it were not cognitive declines, as expected, but *physical,* specifically impaired walking and balance skills. The team of scientists who performed this research had found in a previous study that people who engaged in regular exercise several times a week were up to 40 percent less likely to develop dementia. But the researchers weren't clear about how this relationship worked. "These results [of the new study] suggest that in aging, there's a close link between the mind and body," said Dr. Eric Larson, a scientist who participated in the research. "Physical and mental performance may go hand in hand, and anything you can do to improve one is likely to improve the other."[3] Thus physical activity that includes balance challenges may at least help stall, if not prevent, mental declines. It's another spin on the old adage "use it or lose it," and another penalty to pay for succumbing to "idle-otry."

Of course, for people who are unable to or unwilling to engage in physical activity to keep themselves fit and balanced, technology may come to the rescue within a decade or so. It may be possible, say, for an eighty-year-old snowboarder, who because of age has lost a large percentage of his balance function, to wear a vest-sized vestibular prosthesis under his turtleneck and enjoy the equilibrium of a forty-year-old as he plummets down steep, powdery chutes. Or perhaps scientists will discover a way to control the recently discovered genes that govern vestibular function, allowing us to alter the body's blueprint by regenerating hair cells and thereby permitting the vestibular system to retain its robust sensitivity throughout our long lives.

One of the more intriguing things I discovered about the human balance system is that it is both the oldest and newest of senses. By oldest, I mean on an evolutionary scale. We possess otolith organs, for instance, equivalent in form and function to those of lobsters and crayfish, which originated some 360 million years ago, according to the fossil record.[4] And newest refers of course to how late in the game we humans have come to recognize and understand this enormously important faculty, which only in the past fifty years has begun to achieve what Dr. Terence Cawthorne, the influential British otologist from the 1940s, called "the dignity of a sense."[5] If Aristotle had known what we know today, he would surely have included balance on his list of human senses. Consequently, schoolchildren today would recite six instead of the traditional five. And we would all stand in greater awe of this ancient, intricate, vital, and easily lost power our bodies possess.

Appendix: Balance Exercises

How Good Is Your Balance?

Here's a quick and easy way to evaluate your balance. While standing on a hard surface, raise one leg about a foot off the ground (you may need to have a chair nearby for support). See how long you can maintain this position. Research has shown that for people between the ages of twenty and forty-nine, the average duration is twenty-four to twenty-eight seconds; fifty to fifty-nine years old, twenty-one seconds; sixty to sixty-nine years, ten seconds; seventy to seventy-nine, four seconds. Do this exercise three times and average your performance. This is your baseline figure.

Exercises to Maintain or Improve Your Balance

1. Spend 10 to 15 minutes, three times a week, doing the following sequence of one-legged exercises, either by itself or incorporated into your regular workout. After a couple of weeks, repeat the test and see how your balance improves. As in the baseline balance test, you may want to place a chair next to you for support. As your balance improves (or if it's already pretty good), you can increase the challenge in several ways. You can perform the movements while standing on an unstable

surface (a mattress, cushion, Bosu ball, wobble board, or piece of thick foam). You can move your head slowly back and forth while you're doing the exercises. Or you can try closing your eyes while doing them.

a. In a standing position, with your arms extended straight out to the sides, lift one leg up so that your thigh is parallel to the ground (your lower leg remaining vertical). Hold this position for fifteen seconds. Then lower your leg, rest a few seconds, and repeat five times. Now switch to the other leg and do five repetitions.

b. From the same starting position, raise one leg out to the side, stopping at about forty-five degrees. Hold for fifteen seconds. Switch to the other leg.

c. Extend your left arm straight out in front of you, and your right arm behind, while bending your torso slightly forward. Now raise your straight right leg behind you to a comfortable height and hold for fifteen seconds, repeating five times per side.

d. With your arms out to your sides, standing on one leg, extend the other leg as far as you can, touching imaginary numbers on a clock face (start at noon, say, with your left foot, come back to the center, then proceed to eleven o'clock, and so forth). Bend the knee of your standing leg to reach out farther with your free leg. Switch to the other leg.

2. Heel-to-toe walk. Walk across an imaginary tightrope, placing one foot immediately in front of the other, arms extended to the sides. Now walk backward using the same technique.

3. Look for opportunities to practice standing on one leg. The more time you spend in this position, the better your standing balance becomes. Do it while you're waiting

for the bus, cooking, brushing your teeth, watching TV, or any time you find yourself standing around. Best to resist the urge, perhaps, if you happen to be sharing an elevator with the CEO.

4. Squats and lunges. Leg strength is an important component of balance, whether for elite athletes or older folks getting into and out of a car. These exercises are two of the best for building power in the lower body. If your leg strength is weak, squats can be performed with a chair or bench underneath you for security. Stand in front of it and slowly lower yourself as if you were going to sit down. Just before you contact it, stop and raise yourself slowly up to a standing position. Repeat ten times (or as many times as you can). Younger or stronger people obviously won't require the use of a chair. You can make squats more challenging by holding dumbbells at your sides, or by standing on an unstable surface. To do lunges, stand with arms extended to your sides or hands on your hips. Pretend you're standing in the center of a clock. Extend your left leg out a few feet and touch the "numbers" on the left side of the clock. Bend your stationary right leg as close to the floor as you can. Return to the starting position and repeat with your right leg, touching the numbers on the right side of the clock.

Advanced Exercises

1. Walking the plank. A variation of the heel-to-toe walk, this one uses a long piece of two-by-four lumber, laid on its side, to simulate a tightwire. Find a pen or pencil that has lettering on its side. As you walk slowly along the board, taking small steps, extend the writing instrument in front of you at arm's length, moving it in a

broad figure-eight pattern. Keep your eyes focused on the lettering. This exercise is challenging because you can't use vision to help stabilize you as your gaze is constantly shifting.

(Note: The following exercises are used by the U.S. Ski Team and were provided by team trainer Per Lundstam.)

2. Drop step. Standing on a box or bench twenty to twenty-five inches high, drop onto a soft balance device (Bosu ball, DynaDisc, or thick foam) with both legs. Try to "stick" the landing (remain firmly in position for several seconds) by sinking down smoothly and in a controlled fashion. Repeat several times. An advanced movement is to drop from one leg and land on the same leg again on the balance device. The athlete tries to keep the arms relaxed, in an athletic position, and smoothly contact the balance device.

3. Wall bounce. Facing a concrete wall, you stand in an athletic position on a balance device (rocker board, Bosu ball, DynaDisc, thick foam). Then bounce a ball (weighted medicine ball or basketball) off the wall and catch it. You can bounce the ball off the floor and onto the wall or directly off the wall. For more challenge, stand on one leg. You can also assume a position sideways to the wall and throw the ball across the body.

4. Tug-of-war. Two partners stand on separate balance devices and face each other. Grabbing hold of a rope, they attempt to pull each other off the device, using fakes, feints, and any other sort of treachery they can employ.

5. Weight transfers. Standing with legs wide apart on two balance devices (Bosu balls are ideal), place a broomstick or pole on your shoulders, behind your head. The exercise begins by sinking down on one leg, then slowly performing a weight shift from that leg to the other leg,

keeping the hip travel path as low as possible. Perform the shift slowly, with the lower limb muscles controlling and coordinating it smoothly. Keep the pole level.

6. One-leg floor touches. Stand on the floor on one leg. Imagine small markers placed in a circle about 2 feet from your foot. Drop down and touch the markers around you with one hand, then switch to the other hand. Now switch to standing on the other foot. A further variation is to place your foot on a balance device. The touches should be performed slowly and with optimal control.

Acknowledgments

If there is anyone to credit for setting me on this sometimes stormy but never nauseating voyage on the ship called *Balance,* it is my dad, Bill, though he would be embarrassed to think that it was his falter on Mount Si that started me questioning the nature of balance. Like most people, I had sailed through life without giving balance much of a thought—until the day he fell.

There are many other people to thank as well. Don Parker was kind enough to share his stories and knowledge, and he went beyond the call of duty when he read the manuscript to check for flaws in my understanding and interpretation of the science and physiology of the human balance system. Other scientists who helped shape and clarify my ideas were Dan Merfeld, Paul Bach-y-Rita, Yuri Danilov, Jeffrey Taube, Jim Collins, Ken Erickson, Owen Black, and Larry Rowell.

Frank Belgau gave his time generously, as did Karen Perz, Robin Grindstaff, Cheryl Schiltz, "Crazy Wilson" Dominguez, Gabor Hrisafis, and Colonel (retired) George Maillot.

For her attention to all the minutiae of language, style, punctuation, and logic, I want to thank Little, Brown's sterling copy editor Jayne Yaffe Kemp. For helping me cut away the clutter and keep to the heart of the story, thanks to Liz Nagle, my talented editor. Without the advice and encouragement of my agent, Elizabeth Wales, this book might not have gotten off the ground, much less flown.

Notes

Prologue: In Search of the Lost Sense

1. "A Report of the Task Force on the National Strategic Research Plan," National Institute on Deafness and Other Communication Disorders (NIDCD), National Institutes of Health, April 1989, p. 74.
2. Centers for Disease Control, Fact Sheet: Falls and Hip Fractures Among Older Adults, www.cdc.gov/ncipc/factsheets/falls.htm.
3. Morris, *Wallenda*, pp. 16–17.

Chapter 1: Sickness from Motion

1. Reason and Brand, *Motion Sickness,* p. 187.
2. Ibid.
3. Ibid.
4. Ibid., p. 31.
5. Ibid., p. 188.
6. Money, "Motion Sickness," p. 371.
7. Ibid.
8. Reason and Brand, *Motion Sickness,* p. 3.
9. Ibid., p. 210.
10. Ibid., pp. 7–8.
11. Fisher and Jones, "Vertigo and Seasickness," p. 99.
12. Ibid., p. 101.
13. Thorne, "Cause of Seasickness Discovered at Last?" p. SM12.
14. Reason and Brand, *Motion Sickness,* p. 11.
15. Ibid., p. 16.

16. "We weren't meant to go there," Karadur, www.erowid.org/plants/belladonna.
17. Reason and Brand, *Motion Sickness*, p. 18.
18. Gay and Carliner, "The Prevention and Treatment of Motion Sickness," p. 359.
19. Brinkman, Bill, biography published at www.austincivilwar.org/warstories/bbrinkman.html.
20. Reason and Brand, *Motion Sickness*, p. 21.
21. Ibid., p. 105.
22. Hain, Timothy, "Mal de Débarquement Syndrome," www.dizziness-and-balance.com/disorders/central/mdd.html.
23. Case histories of people who suffer from *mal de débarquement* are posted at www.etete.com/mdd/support.html.
24. Reason and Brand, *Motion Sickness*, p. 71.

Chapter 2: Van Gogh's Ear

1. Hain, Timothy, "Benign Paroxysmal Positional Vertigo," www.dizziness-and-balance.com/disorders/bppv/bppv.html.
2. Ibid.
3. Ibid.

Chapter 3: The Spin Doctors and the Discovery of Multimodality

1. Hawkins and Schacht, "Sketches of Otohistory," p. 185.
2. Cohen, "Erasmus Darwin's Observations," p. 122.
3. Hawkins and Schacht, "Sketches of Otohistory," pp. 185–86.
4. Wade, "Spin Doctors," p. 254.
5. Hunter and Macalpine, *Three Hundred Years of Psychiatry*, p. 596.
6. Ibid., p. 597.
7. Grusser, "J. E. Purkyne's Contributions," p. 136.
8. Bárány, Robert, Nobel lecture, September 11, 1916, posted at www.nobelprize.org.
9. Ibid.
10. Hawkins and Schacht, "Sketches of Otohistory," p. 186.
11. Henn, "E. Mach," p. 145.

12. Ibid., p. 148.
13. Ibid., p. 147.
14. From University of Edinborough Chemistry Department History, posted at www.chem.ed.ac.uk/public/history/history_crumbrown.html.
15. Henn, "E. Mach," p. 147.
16. Finger, *Minds Behind the Brain*, p. 223.
17. Bárány, Robert, Nobel lecture, September 11, 1916, posted at www.nobelprize.org.
18. Jones and Fisher, *Equilibrium*, p. 235.
19. Jones, *Flying Vistas*, p. 87.
20. Jones and Fisher, *Equilibrium*, p. 24.
21. Ibid., p. 25.
22. Jones, *Flying Vistas*, p. 116.
23. Ibid., pp. 111–14.
24. Diamond, "Why Cats Have Nine Lives," p. 586.

Chapter 4: How Balance Contributes to Survival

1. Sylwester, Robert, "We're Inside-Out Crustaceans," May 2002 column for Web site: www.brainconnection.com.
2. Llinás, *I of the Vortex*, pp. 15–18.
3. Luxon, "Vestibular Sytem," in *Vertigo*, p. 1.
4. Ibid., p. 2.
5. Eliot, "Development of Human Ear," *Laryngoscope*, pp. 1160–71.
6. Spoor et al., "Vestibular Evidence," p. 163.
7. Spoor et al., "Implications," p. 646.
8. Bramble, "Endurance Running," pp. 345–52.
9. Krantz, "Brain Size," p. 450.
10. Ibid., p. 451.
11. Darwin, "Origin," p. 418.
12. Loomis et al., "Human Navigation," p. 4.
13. Ibid., p. 5.
14. Maguire et al., "Navigation-related Structural Change," p. 4398.

Chapter 5: "Ear Deaths" and "Graveyard Spirals"

1. Jones, *Flying Vistas*, p. 119.
2. Aeromedical Education Division, "Disorientation."

3. www.iradis.org/education/history/early_history.html (IRADIS is an acronym for International Research and Development Institute for Simulation, a nonprofit organization for vehicle simulation, especially that of airplanes).

4. Keogh, *U.S. Air Mail Service.*

5. Scheck, "Lawrence Sperry."

6. Jones, *Flying Vistas,* pp. 122–23.

7. Personal interview with Don Parker, May 2005.

8. Del Vecchio, *Physiological Aspects,* p. 69.

9. United States Air Force Web site devoted to spatial disorientation: www.spatiald.wpafb.af.mil/index.apx.

10. Ibid.

11. Federal Aviation Administration Advisory Circular entitled "Pilot's Spatial Disorientation," 1983.

12. Of course, as soon as the nineteen pilots in the study began spiraling down, their copilots brought the planes back under control.

13. Alonso-Zaldivar, "Instructor Offered," p. 1.

14. Ibid.

Chapter 6: Tonic and Stimulant

1. Eliot, *What's Going On in There?,* p. 155.

2. Press release from University of Rochester Office of Communications, April 27, 1998, posted at www.rochester.edu/news/show.php?id=185.

3. Clark, "Vestibular Stimulation."

4. Hannaford, *Smart Moves,* pp. 40–41.

5. Although SI treatment isn't embraced by all physicians and psychologists, its theoretical framework has withstood numerous challenges by opponents, and recent brain research has added more credibility to it. It has remained one of the leading forms of treatment for autism and ADD for forty years.

6. Ayres, *Sensory Integration and the Child,* p. 15.

7. Ornitz, "Normal and Pathological Maturation," p. 498.

8. Ayres, *Sensory Integration and the Child,* p. 74.

9. Ratey, *A User's Guide,* p. 307.

10. Hannaford, *Smart Moves,* p. 173.

11. Zero Gravity Corporation's Web address is www.gozerog.com/home_full1.aspx.

Chapter 7: Extreme Equilibrium

1. Croft-Cooke and Cotes, *Circus,* p. 40.
2. Ibid., p. 49.
3. Martin, "Great Alzana."
4. Petyon, "He Just Loves to Scare You," p. 107.
5. Morris, *Wallenda,* pp. 29–34.
6. At least not right away. Years later, other funambulists figured out how to perform similar tricks.
7. Martin, "Great Alzana."

Chapter 8: The Wallenda Within

1. Snow, Anita, "Castro breaks knee, arm in fall after speech," *Seattle Post-Intelligencer,* October 22, 2004.
2. Krucoff, "Before the Fall."
3. CDC, "A Toolkit to Prevent Senior Falls," CDC Web site, www .cdc.gov/ncipc/pubres/toolkit/toolkit.htm.
4. Ibid.
5. National Safety Council Web site: www.nsc.org/issues/fallstop.htm.
6. Ibid.
7. American Academy of Orthopaedic Surgeons Web site: orthoinfo .aaos.org/fact/thr_report.cfm?thread_id=77&topcategory=Hip.
8. Ibid.
9. Hobeika, "Equilibrium."
10. National Vital Statistics Report, vol. 53, no. 5, October 12, 2004, www.cdc.gov/nchs/data/nvsr53/nvsr53_05acc.pdf.
11. National Safety Council Web site: www.nsc.org/library/report_ injury_usa.htm.
12. Forencich, *Play as if Your Life Depends on It,* p. 26.
13. Ibid.
14. Bell et al., "The Road to Obesity," www.obesityresearch.org/cgi/ content/full/10/4/277.
15. Frantzis, *The Big Book of Tai Chi,* p. 2.
16. Ibid., p. 94.
17. Krucoff, "Balancing Acts," p. Z31.
18. Frantzis, *The Big Book of Tai Chi,* p. 96.
19. Robbins et al., "Proprioception and Stability," p. 67.
20. "China's Guangdong Makes 30% of the World's Shoes," *Emerging Markets Economy,* April 8, 2003.

21. Li et al., "Improving Physical Function."
22. Ibid.
23. Forencich, *Play as if Your Life Depends on It,* p. 149.
24. Personal interview with Frank Forencich, November 2005.
25. University of Indiana Media Relations Web site: newsinfo.iu
.edu/tips/page/normal/2379.html.
26. Hobeika, "Equilibrium."
27. Perrin et al., "Effects of physical and sporting activities."
28. Personal interview with Burton Worrell, February 2004.
29. Wolfson et al., "Balance and strength training," p. 498.

Chapter 9: The Cognitive Connection

1. Dr. Erickson's speech can be viewed at www.theblackriver.net/
wobbler/wobblercognitive.html.
2. Bower and Lawrence, "Rethinking the 'Lesser Brain,'" p. 52.
3. Ratey, *A User's Guide,* p. 163.
4. Ibid., p. 157.
5. Blakeslee, "Theory on Human Brain."
6. Ibid.
7. Kiester, "The Sweet Science," p. 32.
8. Leiner and Leiner, "The Treasure."
9. International Dyslexia Association Press Release, October 23,
2003, entitled "Controversial Therapy Lacks Research Basis."
10. Reprint of Reynolds et al., "Evaluation of an exercise-based
treatment," p. 9.
11. Kiester, "The Sweet Science," p. 32.

Chapter 10: New Balance

1. Bach-y-Rita et al., "Vision Substitution," p. 964.
2. Blakeslee, "New Tools," p. D1.
3. Priplata et al., "Noise-Enhanced Balance Control."
4. "A Report of the Task Force on the National Strategic Research
Plan," National Institute on Deafness and Other Communication Disorders (NIDCD), National Institutes of Health, April
1989, p. 39.
5. Rubel, "Promising Research."
6. Travis, "Getting an Earful."

Epilogue: Staying Glued to the Wire

1. Bles et al., "The mechanism of physiological height vertigo."
2. Wang et al., "Performance-based physical function."
3. Group Health Cooperative News Release, May 22, 2006: www.ghc.org/news/news.jhtml?reposid=/common/news/news/20060522-dementia.html.
4. Lobster age data provided by Dr. David A. Kendall, an entomologist, on his Web site: www.kendall-bioresearch.co.uk/crust.htm#fossil.
5. Cawthorne, "Vestibular Injuries."

Bibliography

Aeromedical Education Division. "Disorientation, or Whose Gyros Can You Trust?" Brochure created by the Civil Aerospace Medical Institute, Federal Aviation Administration. Publication: AM-300-90/1.

Alonso-Zaldivar, Ricardo. "Instructor Offered to Fly with JFK Jr., Report Says." *Los Angeles Times,* July 7, 2000, p. 1.

Ayres, A. Jean. *Sensory Integration and the Child.* Los Angeles: Western Psychological Services, 1979.

Bach-y-Rita, Paul, Carter Collins, Frank Saunders, Benjamin White, and Lawrence Scadden. "Vision Substitution by Tactile Image Projection." *Nature* 221 (1969): 963–64.

Bell, A. Colin, Keyou Ge, and Barry Popkin. "The Road to Obesity or the Path to Prevention: Motorized Transportation and Obesity in China." *Obesity Research* 10 (2002): 277–83.

Berthoz, Alain. *The Brain's Sense of Movement.* Cambridge, Mass., and London, England: Harvard University Press, 1997.

Blakeslee, Sandra. "New Tools to Help Patients Reclaim Damaged Senses." *New York Times,* November 23, 2004, p. D1.

———. "Theory on Human Brain Hints How Its Unique Traits Arose." *New York Times,* November 8, 1994, p. C1.

Bles, W., T. S. Kapteyn, T. Brandt, and F. Arnold. "The mechanism of physiological height vertigo." *Acta Otolaryngol* 89, nos. 5–6 (May–June 1980): 534–40.

Bower, James, and Lawrence Parsons. "Rethinking the 'Lesser Brain.'" *Scientific American,* August 2003, pp. 51–57.

Bramble, Dennis, and Daniel Lieberman. "Endurance Running and the Evolution of Homo." *Nature* 432 (2004): 345–52.

Cawthorne, Terence. "Vestibular Injuries." *Proceedings of the Royal Society of Medicine,* vol. 39, November 1945–October 1946, p. 273.

Clark, D. L., et al. "Vestibular Stimulation Influence on Motor Development in Infants." *Science* 196 (1977): 1228–29.

Cohen, B. "Erasmus Darwin's Observations on Rotation and Vertigo." *Human Neurobiology* 3 (1984): 121–28.

Croft-Cooke, Rupert, and Peter Cotes. *Circus: A World History.* New York: Macmillan, 1977.

Darwin, Charles. "Origin of Certain Instincts," *Nature* 179 (1887): 417–18.

Del Vecchio, Robert J. *Physiological Aspects of Flight.* Oakdale, N.Y.: Dowling College Press, 1977.

Diamond, Jared. "Why Cats Have Nine Lives." *Nature* 332 (April 14, 1988): 586–87.

Eliot, Lise. *What's Going On in There? How the Brain and Mind Develop in the First Five Years of Life.* New York: Bantam, 1999.

Finger, Stanley. *Minds behind the Brain: A History of the Pioneers and Their Discoveries.* New York: Oxford University Press, 1999.

Fisher, Lewis, and Isaac Jones. "Vertigo and Seasickness: Their Relation to the Ear." *New York Medical Journal,* July 15, 1916.

Forencich, Frank. *Play as if Your Life Depends on It: Functional Exercise and Living for Homo Sapiens.* Go Animal, 2003.

Frantzis, Bruce. *The Big Book of Tai Chi: Build Health Fast in Slow Motion.* London: Thorsons, 2003.

Gay, Leslie, and Paul Carliner. "The Prevention and Treatment of Motion Sickness." *Science* 109 (April 8, 1949): 359.

Grusser, O. J. "J. E. Purkyne's Contributions to the Physiology of the Visual, the Vestibular and the Oculomotor Systems." *Human Neurobiology* 3 (1984): 129–44.

Hannaford, Carla. *Smart Moves: Why Learning Is Not All in Your Head.* Salt Lake City: Great River Books, 2005.

Hawkins, Joseph, and Jochen Schacht. "Sketches of Otohistory, Part 8: The Emergence of Vestibular Science." *Audiology Neurotology* 10 (2005): 185–90.

Henn, V. "E. Mach on the Analysis of Motion Sensation." *Human Neurobiology* 3 (1984): 145–48.

Hobeika, Claude. "Equilibrium and Balance in the Elderly." *Ear Nose and Throat Journal,* August 1999.

Hunter, Richard, and Ida Macalpine, eds. *Three Hundred Years of Psychiatry, 1535–1860: A History Presented in Selected English Texts.* Hartsdale, N.Y.: Carlisle Publishing, 1982.

James, William. *Essays in Psychology.* Cambridge, Mass., and London: Harvard University Press, 1983.

Jones, Isaac, and Lewis Fisher. *Equilibrium and Vertigo.* Philadelphia: Lippincott, 1918.

Jones, Isaac. *Flying Vistas: The Human Being Through the Eyes of a Flight Surgeon.* Philadelphia and London: Lippincott, 1937.

Keogh, Edward. *Saga of the U.S. Air Mail Service,* 1927, as excerpted at www.airmailpioneers.org/history/Sagahistory.htm.

Kiester, Edwin, Jr. "The Sweet Science of Movement," *Pitt Med,* July 2002, pp. 29–32.

Krantz, Grover. "Brain Size and Hunting Ability in Earliest Man." *Current Anthropology 9* (1968): 450–51.

Krucoff, Carol. "Balancing Acts: How Exercise Can Help Seniors Prevent Falls." *Washington Post,* April 30, 1996, p. WH31.

———. "Before the Fall: Dancing, Strength Training and Other Exercises May Stem the Rise in Seniors' Injuries." *Washington Post,* July 25, 2000, p. Z8.

Leiner, Henrietta, and Alan Leiner. "The Treasure at the Bottom of the Brain." www.newhorizons.org/neuro/leiner.htm.

Li, Fuzhong, K. Fisher, and Peter Harmer. "Improving Physical Function and Blood Pressure in Older Adults Through Cobblestone Mat Walking: A Randomized Trial." *Journal of the American Geriatrics Society 53* (2005): 1305–12.

Llinás, Rodolfo. *I of the Vortex: From Neurons to Self.* Cambridge, Mass., and London: MIT Press, 2001.

Loomis, Jack, Roberta Klatzky, Reginald Golledge, and John Philbeck. "Human Navigation by Path Integration." In *Wayfinding: Cognitive mapping and spatial behavior.* Edited by R. G. Golledge. Baltimore: Johns Hopkins Press, 1997.

Luxon, Linda. "The Anatomy and Physiology of the Vestibular System." In *Vertigo.* Edited by M. Dix and J. Hood. New York: John Wiley, 1984.

Maguire, Eleanor, David Gadian, Ingrid Johnsrude, Catriona Good, John Ashburner, Richard Frackowiak, and Christopher Frith. "Navigation-related Structural Change in the Hippocampi of

Taxi Drivers." *Proceedings of the National Academy of Sciences* 97 (2000): 4398–403.

Martin, Douglas. "Great Alzana, who mocked death and New York law on the high wire, dies at 82." *New York Times,* March 18, 2001.

Money, K. E., and W. S. Myles. "Motion Sickness and Other Vestibulo-Gastric Illnesses." In *The Vestibular System: The proceedings of a symposium held at the University of Chicago, 1973.* Edited by Ralph Naunton. New York: Academic Press, 1975.

Morris, Ron. *Wallenda: A Biography of Karl Wallenda.* Chatham, N.Y.: Sagarin Press, 1976.

Ornitz, Edward. "Normal and Pathological Maturation of Vestibular Function in the Human Child." In *Development of Auditory and Vestibular Systems.* Edited by R. Romand. New York: Academic Press, 1983.

Perrin, P. P., G. C. Gauchard, and C. Jeandel. "Effects of balance and sporting activities on balance control in elderly people." *British Journal of Sports Medicine* 33 (1999): 121–26.

Peyton, Bernard, Jr. "He Just Loves to Scare You." *Saturday Evening Post,* August 12, 1950, pp. 105–10.

Priplata, Attila, et al. "Noise-Enhanced Balance Control in Patients with Diabetes and Patients with Stroke." *Annals of Neurology* 59 (2006): 1123–24.

Ratey, John J. *A User's Guide to the Brain: Perception, Attention, and the Four Theaters of the Brain.* New York: Pantheon, 2001.

Reason, J. T., and J. J. Brand. *Motion Sickness.* London and New York: Academic Press, 1975.

Reynolds, David, Roderick Nicolson, and Helen Hambly. "Evaluation of an exercise-based treatment for children with reading difficulties." *Dyslexia,* February 2003.

Robbins, Steven, Edward Waked, and Jacqueline McClaran. "Proprioception and stability: Foot position awareness as a function of age and footwear." *Age and Ageing* 24 (1995): 67.

Rubel, Edwin. "Promising Research on Hair Cell Regeneration." *The Hearing Review,* October 2004. www.hearingreview.com/article.php?s=HR/2004/10&p=1.

Scheck, William. "Lawrence Sperry: Genius on Autopilot." *Aviation History Magazine,* November 2004.

Spoor, F., S. Bajpai, S. Hussain, K. Kumar, and J. Thewissen. "Vestibular Evidence for the Evolution of Aquatic Behaviour in Early Cetaceans." *Nature* 417 (2002): 163–65.

Spoor, F., B. Wood, and F. Zonneveld. "Implications of Early Hominid Labyrinthine Morphology for Evolution of Human Bipedal Locomotion." *Nature* 369 (1994): 645–48.

Thorne, Van Buren. "Cause of Seasickness Discovered at Last?" *New York Times,* October 29, 1916, p. SM12.

Travis, John. "Getting an Earful: With Gene Therapy, Ears Grow New Sensory Cells." *Science News Online* 163 (2003).

Wade, Nicholas. "The Original Spin Doctors—The Meeting of Perception and Insanity." *Perception* 34 (2005): 253–60.

Wang, L., E. B. Larson, J. D. Bowen, and G. van Belle. "Performance-based physical function and future dementia in older people." *Archives of Internal Medicine* 166, no. 10 (May 22, 2006): 1115–20.

Wehner, R., and S. Wehner. "Path integration in desert ants: Approaching a long-standing puzzle in insect navigation." *Monitore Zoologico Italiano* 20 (1986): 309–31.

Wolfson, L., R. Whipple, C. Derby, J. Judge, M. King, P. Amerman, J. Schmidt, and D. Smyers. "Balance and strength training in older adults: intervention gains and Tai Chi maintenance." *Journal of the American Geriatric Society* 44 (1996): 498–506.

Index

CPSIA information can be obtained at www.ICGtesting.com
Printed in the USA
LVOW040927070912

297774LV00002B/1/P